AF678402

DOG

KILL

Books by Al Dempsey

Phase of Darkness

Dog Kill

DOG

KILL

by Al Dempsey

PRENTICE-HALL, INC., Englewood Cliffs, New Jersey

DOG KILL

by Al Dempsey

Copyright © 1976 by The Moorings of Nassau, Ltd., Inc.

All rights reserved. No part of this book may be reproduced in any form or by any means, except for the inclusion of brief quotations in a review, without permission in writing from the publisher.

Printed in the United States of America

Prentice-Hall International, Inc., London
Prentice-Hall of Australia, Pty. Ltd., Sydney
Prentice-Hall of Canada, Ltd., Toronto
Prentice-Hall of India Private Ltd., New Delhi
Prentice-Hall of Japan, Inc., Tokyo

Library of Congress Cataloging in Publication Data
Dempsey, Al.
Dog kill.

I. Title.
PZ4.D3848Do3 [PR6054.E47] 823'.9'14

ISBN 0-13-217257-7 75-46552

10 9 8 7 6 5 4 3 2

For Paddy

A message from the American Humane Association AHA

In this novel many of the most shocking scenes involving wild dogs are, tragically, based on actual incidents. As Mr. Dempsey ably shows, feral and uncontrolled dogs *are* a growing national problem. In certain circumstances they can be dangerous to humans, and they have undoubtedly accounted for tens of millions of dollars worth of losses to our wildlife, livestock, and poultry resources.

There is no contradiction between admitting these harsh facts and continuing to love and respect dogs. Nor can there be a question of blaming any dog for behaving like a wild animal if it has been condemned to *be* a wild animal. Indeed, it is important to understand where the blame really does lie: It lies not with dogs, but with people. It lies with those owners who refuse to accept responsibility for controlling their pets; who fail to take action to prevent unwanted litters; and who, through callousness, cruelty, or neglect, contrive to turn man's best friend into an enemy.

Don Hegg
Associate Director, Information Services
American Humane Association
National Headquarters
Denver, Colorado

The American Humane Association is a national non-profit federation established in 1877 to prevent cruelty to children and animals.

DOG

KILL

Chapter One

Mel Ryan stared through the windshield, trying to ignore the gnawing sensation in the pit of his stomach. The two-way radio in his pickup truck crackled again, fighting for his attention, and he reached for the microphone. The plastic was cold to his touch as he pressed the transmit button and said, "Roger, Base. This is Unit Twelve."

The Vail police radio operator asked, "How long you gonna be? Your office is on the phone and they want to know."

Mel paused, gauging the rough fifteen miles back to Vail. "I should be there soon. Tell them about an hour."

The radio was silent for a few seconds while the message was relayed, then word came back: "Okay, but they've got a bitch in heat over at the Red Lion Condo and it seems every dog in town has picked up the scent."

Mel acknowledged the message and replaced the mike, then went back to watching.

The cab of his Animal Control pickup was still warm, even though he had turned off the engine several minutes before. The vehicle was in snow up to its running board, which was just over two feet off the roadbed. The logging road to Timber Tract 734 hadn't been used since the first snow and the accumulation was nearly enough to stop him, even with his powerful four-wheel drive and snow chains on his studded tires.

He looked through his windshield in the direction of the right side of the roadbed and studied the old bull elk again. There was no doubt about it: The panting, snorting animal was not only stuck in a snowdrift that covered his back with white powder, but he was on the edge of exhaustion. Mel's eyes moved back in the opposite direction, across the road, and up to a ledge of rock that had been blown nearly clear of snow; and, once again, he felt a pull in his stomach.

There were only three of them: a docile-looking Saint Bernard, a dog of mixed breed that was probably mostly pointer, and another whose parentage had been dominantly German shepherd, though with a strong Irish setter somewhere in the woodpile. It was the shepherd that was causing the tight feeling in Mel's stomach. While the other two dogs were intent on watching the snow-captured elk, the shepherd was watching Mel.

Mel had worked professionally with dogs since his days in pre-veterinarian school, and when he had been forced to leave school, he had chosen a job as an animal control officer in the city of Boston. In that position he had learned to handle dogs without the use of either the noosed "dog stick" or the tranquilizer gun. Mel insisted on "talking" to a stray, no matter how nasty the animal seemed to be; he was convinced that the scent of fear never manifested itself on him, and so he had never been attacked.

But fear clutched at him now. The shepherd was poised, watching Mel intently, and even with the truck's door closed, Mel was convinced that the shepherd was somehow aware of his fear.

A bellow from the elk snapped Mel's attention back to the snow-filled gully. The old bull was getting into more trouble as he struggled in the deep drift. He had slipped to his front knees, and his head was thrashing back and forth in an effort to get free.

Mel felt the electricity of movement before a blur slid into his peripheral vision; it was the dark brown head and spotted white body of the pointer as it lunged down from the rock ledge toward the elk. Mel had hoped that the presence of the truck would discourage the dogs, but the hope vanished as he saw the pointer land in the snow and make for the trail that had been opened by the elk. As Mel debated his next move, he noticed the huge Saint Bernard tensing to leap and follow the pointer. He looked for the shepherd, and his heart thumped. The dog had left the other two and was moving toward the truck, its dark, penetrating eyes fixed on Mel.

The use of his handling stick was now out of the question with the other two dogs so near, and the tranquilizer gun was a single-shot weapon and therefore useless if the elk was to be saved. His options were narrowed to one—his rifle. He hated to shoot a dog, but the elk was important, too, and the law called for destroying any dog caught in the act of ravishing wildlife. The year before, in Colorado alone more than thirty elks had been identified as killed by wild dogs, a dangerous loss to the Rocky Mountains' wildlife population.

He lifted the Remington 30-06 from the gun rack in back of his head and opened the door. Quickly he moved his foot out onto the running board. As he actioned a round into the chamber and flicked off the safety, he saw that the shepherd was increasing its speed, its pads spread wide for traction in the snow.

Mel's original intention had been to shoot at the pointer, thus halting the attack on the elk and scaring off the other dogs with the noise of the gun. But, as he raised himself into a standing position on the running board to get off the shot, he realized the shepherd was actually going to attack the truck. He arched his left arm over the top of the open truck door and dropped the front of his rifle into his left hand, but even as he brought the stock to his shoulder he realized he had underestimated the shepherd's ability to move through the thick snow. At first, as Mel struggled to sight down the barrel, he thought the dog was making a futile effort to clear the top of the door. But as he felt the weight of the door slam into his chest, Mel realized that the dog had done just what he wanted to do—throw a body check into the door. The dog landed in the soft snow beside the truck, and Mel slipped back onto the seat of the truck, the rifle jammed over the top of the now partially closed door. As he struggled to free the gun, he felt the burning, tearing pain of fangs penetrating the calf of his leg. For the first time in his life, he had been bitten by a dog.

Fighting against pain and panic, he looked down toward the gully, where the elk was bellowing in agony. He saw that the pointer had hit the old bull's right rear flank and ripped a flap of fur, skin, and muscle down to the hipbone. The Saint Bernard, as if performing some bizarre circus act, was trying to balance on the elk's back.

Mel grabbed the rifle stock with his right hand and

the door handle with his left, and he had nearly cleared the rifle into the truck's cab when he felt the pulling on his leg stop. The relief was only momentary, because the shepherd had let go only to lunge again, this time for Mel's left wrist. Mel anchored his right foot to the brake pedal and kicked at the shepherd's underside with his aching left leg. He flipped the rifle onto the floor of the truck and grabbed at the top of the door with his free right hand. The snarling of the dog and the bellowing of the elk filled Mel's ears, and he felt the warm blood from his new wounds mingle with the shepherd's saliva. He kicked again, but the yelp of pain did not force the tenacious dog to let go. Struggling against the pull of the heavy, thrashing body, he strained painfully to bring the dog's head in line with the jamb of the truck's door, then yanked inward with all his strength. Two of his fingers broke as the cold metal frame banged at the top of the door, but down below, the blow jolted the dog enough to make it release Mel's wrist. He slammed the door shut.

Panting, Mel looked back toward the gully, where the elk was still bellowing. The Saint Bernard was tearing at the back of the animal's neck, its heavy fur protecting it from the defensive movement of the elk's antlers.

Outside the cab of the truck, the shepherd had flattened out an area in the snow so that he was able to propel himself repeatedly into the air and was slashing viciously at the glass windows in an attempt to get at Mel. First the dog's large pads and long claws would come into view as it struck violently against the truck. Then the head would appear, flews flapping and lips pulled back, exposing blood-flecked teeth. Then the scene would disappear and give way to a pastoral view of a snowy forest framed in the truck's window, marred only by the scrim of liquid oozing in the foreground.

Mel stretched out across the seat of the truck and pulled the rifle up so it was pointed at the window. The shepherd rose into view and Mel pulled the trigger. The glass of the window shattered into a brilliant spiderweb, and the dog vanished.

Mel pushed the door open cautiously. The shepherd was in a snowbank, still twitching, its head the center of a spreading crimson circle.

The pain from his leg and arm wounds made Mel light-headed, but the scene down in the gully almost made him vomit.

The battle between the dogs and elk had flattened out a small area of the snow, so the elk could have moved—if he had been able to move. But the pointer, having ripped the tendons in the elk's left rear hip, was now tearing open the wapiti's belly. The elk could not defend itself because its antlers were jammed and lodged in the throat of the Saint Bernard.

Leaning against the side of the truck, Mel pulled the gun up to his cheek and quickly fired a round that hit the Saint Bernard just below and in back of the shoulder and into the heart. The death spasm shook the dog loose from the elk which pulled its head back and up and gave another painful cry as it tried to reach the pointer.

Mel did not have a clear shot at the pointer so he fired into the snow about five feet in back of the elk. The pointer bolted free, and Mel saw the elk try to lunge after the remaining dog, then collapse on its torn leg.

The pointer ran back up the broken trail toward the Timber Road and Mel pulled the gun back to his cheek.

Suddenly, without apparent cause, the pointer stopped dead in his tracks and looked in Mel's direction. Mel lined the dog in the sights and began to tighten his finger on the trigger. The pointer stood there in the snow, not moving except for the heavy breathing that caused puffs of steam to come from his mouth. Mel held the gun at his cheek, but relaxed the tension on the trigger and studied the animal. He remembered a time back in his youth when his father had taken him hunting with their own pointer, and he realized that this dog was like a snapshot from his own past—except for the blood dripping from its mouth and splotching its coat. Mel felt a strange sense of anger well up inside him, and he began to once again tighten his grip on the trigger.

The dog seemed to sense the threat and flattened itself on its belly. And then, amazingly from deep down in that *other* part of a dog's heart, the pointer offered a furious wag of his tail, and he eased his head forward with a slight twist that has been, for thousands of years, the canine reaction to "good boy." Mel yanked the gun down, unable to pull the trigger.

The two of them stood there for nearly a minute, the pointer wagging his tail tentatively and Mel debating with his own good judgment. Then Mel stooped down, snapped his fingers, and ordered, "Come!"

The dog convulsed in a fit of wagging, but he did not move from his spot.

"Come!" Mel repeated sternly.

The pointer dropped down on his forelegs like a puppy, his tail flipping back and forth at the back of his elevated rump.

"Come," Mel pleaded in a more gentle tone.

The pointer gave one low, happy, friendly Grruuufh

and bounded off back up the trail toward the rock ledge. Mel didn't bother to bring his rifle up again; he had already made his decision. He watched as the pointer made an easy five-foot leap up onto the ledge, and in an instant was gone from sight.

A long, low moan pulled Mel's attention back to the gully. The elk had dragged himself away from the battle area and was once again getting trapped on the soft, powdery snow. Mel worked his way down through the deep white drifts and came first to the Saint Bernard, sprawled in an area carpeted with blood. Mel bent down and looked at the dead dog.

"Shit," he said as he slid his hand into the thick hair at the neck and pulled at the collar hidden in the fur. There was an Eagle County dog registration tag hanging from it, clear evidence that this was no wild dog but someone's pet, a pet whose owners were willing to pay the three dollars fifty cents a year to have him licensed.

Another groan brought Mel's attention back to the blood-red snowdrift where the elk had pulled himself. Mel walked over and did not need more than a few seconds to make his decision on this animal. With a ruined leg and his bowels torn open, it would only be a matter of hours—painful and inhumane hours—before the elk expired or was attacked again by some animals. One shot ended the elk's obvious agony.

Off in the distance, far up on the side of the tree-covered hills, a dog barked. Mel knew the sound. It wasn't a wild bark; it was the response a hunting dog gives when his master has made a kill. Mel looked up toward the dense woods in the direction the pointer had fled. The bark came again and Mel shouted in frustration, "Oh, go to hell!"

Chapter Two

AHA AMERICAN HUMANE ASSOCIATION

ANIMAL CONTROL STUDY

Computer Control Center, Denver, Colorado

EXHIBIT NUMBER 26-5-78

DOG PARTICULARS

BREED Harrier

COLOR Brown, black, white

WEIGHT 35 lbs. HEIGHT app. 18"

NAME (IF KNOWN) Blackie

AGE (APPROX.) 1 year

ANIMAL CONTROL REGISTRATION

Lowenthal Research Center, University of Indiana

CITY (COUNTY) STATE NUMBER

TYPE OF LOSS Release from research lab.

Mary Ellen Crawford looked anxiously at her watch, then across the room at Reggie. He was sitting there, studying Mary Ellen with an intensity that was familiar but still disconcerting—she had always felt uncomfortable about the way he looked at her.

Reggie nodded his head slightly, then made a sideways motion in the direction of the window. Mary Ellen was grateful for the release. She eased herself off the typing chair, walked around the desk, and went to look out of the window.

She knew what she would see because they had rehearsed the entire operation a half-dozen times over the past week: Shadows were lengthening across the campus; auto traffic was light on South Jordan Avenue; few students were in view. The Indiana University campus in Bloomington was hibernating during the long Thanksgiving vacation.

When Mary Ellen had first become involved in the project, she had felt some guilt about using her position of trust here at the International Center. But Reggie had convinced her that it was vital for them to have access to this building, and they had come here every night for the past week to check on schedules and observe routines. But this night was different: Tonight they would go through with Reggie's plan, and the qualms were returning. She knew it would be easy to abort the operation by merely walking out of the building and leaving Reggie alone. Still, deep down, she believed what they were doing was right. The director of the International Center had trusted her with the key to the building over the holidays, and she did not like abusing the good faith. But it was too late now; she was committed.

She stared up Jordan Avenue and saw the blue and white university police car emerging from Seventh Street. It turned right and drove slowly south on Jordan.

Mary Ellen took a deep breath.

"They're coming," she said quietly. "Right on time."

Reggie sprang from the sofa and was at the telephone in two long strides. He dialed a number.

Mary Ellen knew what the train of events would be from that moment on: The phone would ring in the Book Nook over on East Kirkwood, then Dolly and Clark would walk across the street and get into their Ford Econovan and start driving. The carefully timed trip would require exactly eighteen minutes.

Mary Ellen kept looking down at the police car as it came toward the International Center. She knew they could not see her up in the window on the third floor, and she almost wished they could. She felt a strong temptation to wave her arms and attract their attention.

Behind her, she heard Reggie's voice saying, "Okay, Dolly. Let's get this going." The telephone dropped back into its cradle, and she could feel him approaching her from across the room.

Down on Jordan Avenue, the campus police car had stopped in front of the Loventhal Medical Research Building; that was routine. One of the policemen climbed out and went to make sure the front door was locked; still the routine held.

She was suddenly aware that Reggie had eased up behind her. His body touched hers as he looked out the window over her shoulder, and she felt his warm breath on her ear. She saw their reflections on the windowpane and wondered again why he had never made any physical move toward her. They had been alone in that same office every night for a week, a couple of hours each time, yet he had never touched her.

"They're sure taking their time," Reggie whispered, as if his words could be heard by the policemen across the street.

The policeman checked the front door, then walked around the side of the building and checked that the truck-delivery entrance was locked. The security inspection completed, he got back into the car. In a moment the car began to move off.

Reggie nodded. "Okay, get at it."

Obediently, Mary Ellen went to the telephone and dialed the number she had memorized. In a moment, she said into the phone, "Hi, Jimmie?" There was a pause, then she continued, "You don't know me, but I'm doing some typing for Dr. Spina. He's away, you know, but he just called me and asked me to pick up some papers. He said you were on duty tonight and you'd let me into the building." Another pause, and she smiled. "I promise," she said. "I'll be at the front door in a couple of minutes."

She hung up and nodded to Reggie.

A thin grin came to his lips and he pointed at the door. "Okay. Now we show those sonsofbitches."

Mary Ellen left the office, nearly ran down the stairs to the ground floor, and went out into the cold fall evening. She walked quickly toward the seven-story Research Building. The only lights burning were in the entrance lobby and up on the sixth floor. She stopped again in front of the building, staring upward.

This was the first time Mary Ellen Crawford had ever done anything really illegal in her life. She had smoked pot, of course, and broken a few traffic rules in her Mustang, but this was different. It was frightening, but it was important, too. She was certain of that.

The Loventhal Medical Research Center was a recent addition to Indiana University, built with a grant from philanthropist Joseph Loventhal coupled with federal and state funds. The center operated through research monies from the state medical association, private contributors, corporate endowments, and Department of Defense project contracts. The latter was the cause of much campus turmoil. It had been impossible to keep the nature of the DOD experiments secret, and by now the entire student body knew that up on the sixth floor of the Loventhal Research Building an elaborate vivisection laboratory was conducting military-oriented research.

To Mary Ellen, using dogs in research was bad enough, but when the word leaked out about the DOD project, her reaction was nearly violent. It did not take her long to find kindred spirits in Reggie and his friends Dolly and Clark. With a minimum of investigation, they found out exactly what was going on.

They learned that there were twenty-seven dogs up on the sixth floor, being used in a wide variety of experiments. One involved Teflon clips, which might lead to the replacement of sutures in up to forty percent of the surgical cases now needing stitches. Four of the pups had embedded in them up to twenty clips now under study, another group of dogs was being used in coronary arterial bypasses, and another collection of pups was being operated on every other day to discover the extended effects of electrothermal anesthesiology, a technique that could lead to surgery on humans without the side effects of present anesthesia. Two dozen dogs had already died on the operating table.

Mary Ellen was too intelligent to discount the value of vivisection, and she knew that most of the dogs involved in experiments came from animal shelters and were destined to be destroyed because communities could not afford to support stray dogs on a permanent basis. She admitted that it made sense to use these dogs for the benefit of mankind, just as long as the work was done humanely.

However, Reggie had done some digging into what was going on, and he had found out that the laboratory had a separate section set aside as part of a Department of Defense medical research grant to explore the effects of laser surgery on battlefield injuries. The problem was that the dogs had to be inflicted with battlefield-type wounds to see how the laser instruments could be applied to combat operations. Reggie had found out that three of the dogs had been subjected to napalm burns; five of them had had sundry types of head injuries induced to represent shrapnel wounds; two had had their pelvic regions crushed, as if a tank or truck had run over them; and five of the dogs had their stomachs slit open to simulate bayonet injuries. The word was that the attending dog handlers made it a point to stay away from that area of the kennel.

The memories of Reggie's graphic description still caused a queasy feeling inside Mary Ellen even now, as she walked toward the entrance to the building. As she approached the glass door, she could see the vivisection lab attendant coming out of the elevator. He offered a friendly wave of the hand, and she answered with a big smile and a shrug of her shoulders that indicated she was sorry to bother him.

He opened the door, and she again apologized as they walked across the lobby and into the elevator. As the door closed, Mary Ellen reached across and touched the fourth-floor button.

"Hey," he said, "Dr. Spina's office is on the sixth floor."

"I know but I've got to get some papers from Dr. Lippman's first. I'll come up to the sixth in a few minutes. Okay? Do you have any hot coffee up in the lab? It's freezing outside."

The young man knew it wasn't freezing yet, even if

the temperature had begun to drop since the sun started setting. He looked at her with newly formed anticipation. She was a pretty girl. Her long blond hair framed her sculptured face perfectly. She wore a bulky Irish-knit sweater and tight-fitting Levis, neither of which concealed her inviting body.

He began to run through a mental checklist: Put coffee cups in the reception room by the couch; wash out his mouth with Lavoris; turn the lights down; give the dogs biscuits so they'd shut up for a while. The rest of the scene roared through his imagination like a train in a tunnel as he watched her get out of the elevator.

Mary Ellen turned and smiled encouragingly. "Coffee?"

He cleared his throat. "Hell, yes," he said. "I've got gallons of it. Come on up."

"Great," she said encouragingly as the door slid closed quietly.

She waited in the dimly lit hall for a minute or two, then pushed the down button. The elevator appeared in seconds. Once back down in the lobby, she scooted around through a side door and made her way back to the incoming parcel area. Ingenuously, Reggie had found out the exact location of everything concerned with the operation, and Mary Ellen was able to go right to the green and red buttons that operated the delivery door. Before it was fully up, Clark was backing the Econovan into the truck bay and Reggie was jumping out.

"Close it!" His voice was harsh. "Push the frigging close switch."

Mary Ellen fumbled for a moment but finally hit the switch, and the door rumbled back down.

In the darkness of the big room, Dolly had slipped

out of the truck and was at Mary Ellen's side, whispering excitedly, "You did it! You did it! We did it!"

"Let's knock it off," Reggie said sternly. "We've got work to do."

Clark opened the double doors of the truck from the inside. He stepped out, waving a large army fatigue jacket, the sleeves of which had been sewn closed and extended with long pieces of canvas to make it serve as a straightjacket.

"Let's go," Reggie ordered, and the four students entered the elevator. At the sixth floor, Mary Ellen called to the kennel attendant from the hall. When he came out of the lab, Reggie and Clark jumped him and forced him into the fatigue jacket, tying the sleeves behind him. They dragged him to the elevator and dispatched him down to the first floor. Then, as they had planned, Reggie led Mary Ellen and Dolly to the cage area and began releasing the test dogs from their cages.

He left the girls loading dogs onto the freight elevator and returned to the lab to destroy any research files he could find.

While Clark was smashing microscopes, centrifuges, and other lab paraphernalia, Reggie began dumping file folders into a large, stainless-steel tub used to bathe the dogs. He was pouring a thick, deep-purple duplicating-machine fluid into the tub full of research papers when a scream came from the kennel area.

Reggie stiffened, then ordered Clark to continue. "I'll see what the hell they're up to," he said.

Reggie ran through the door and saw Dolly fighting with a small dog that had obviously bitten her. Mary Ellen was running out the door into the hall after a larger dog.

He yelled, "Damn it, Dolly. Get another dog. I'll take

this one." He scooped up the small dog and moved quickly out into the hall, where ten or eleven dogs milled around in confusion. He herded them inside the freight elevator and pulled up the wooden safety gate, which would act as a containing barrier. He ran back down the hall and found Mary Ellen trying to coax a dog to come with her.

Reggie shoved her aside roughly and grabbed the dog by the nape of the neck. He flipped it up into his arms and spun around, bumping into Mary Ellen.

"Get the hell out of my way," he said angrily.

The dog in Reggie's arms whimpered, and Mary Ellen pleaded, "Be gentle with him."

Reggie glared. "Look, you worthless shit. I only brought you because you could get us in here. We don't need you now, so stay out of my way."

"But the dog—"

"Screw the dog!" he shouted. "We're here to show these tin-hat bastards they can't run our school." He pushed past her.

Mary Ellen stood there, staring in confusion. She had joined the dog rescue because she felt the torture of animals in the name of research was wrong. She was sure Dolly and Clark felt the same way. She ran after Reggie.

By the time she caught up with him, Dolly was coming out of the kennel area with another yelping dog, and Reggie was trying to get still another one into the freight elevator.

At that moment the building shook, and smoke billowed from the laboratory into the hall.

Reggie shouted, "Stay with the dogs." He ran back through the kennel area and into the operating room. He found Clark leaning against a wall, dazed.

Half of the room was in flames. Cans of ether were

exploding, an oxygen outlet had broken off and was burning like a blowtorch, and chemicals were beginning to ignite or fume. Clark's beard, eyebrows, and forelocks had been singed off, and flash burns covered his face and forearms.

"You Goddamned idiot!" Reggie screamed and slammed a fist into Clark's shoulder. He pulled Clark into the kennel room and out into the hallway. Dolly was standing there screaming, and Reggie did the only thing he could think of: He shoved both of them into the elevator with the dogs. He turned to Mary Ellen, who was chasing a dog down the hallway, and yelled, "Come on! We're leaving!"

Mary Ellen whirled and stared at him. "We can't leave the dogs. They'll be killed. My God, Reggie, the fire!"

Reggie took two strides and grabbed her by the shoulders. "Look, Miss Rich-Bitch, those dogs aren't going to burn. The fire will be out in a few minutes. I never wanted you along in the first place, so shut your damned mouth and get moving. It's people like your old man who support this sort of crap."

She was yelling now. "You stupid jerk. What's that got to do with anything? Those dogs in there are in danger. We have no right to just leave them!"

Reggie flung her away angrily. She stumbled and fell into the back of the large elevator, right down beside the milling dogs. He followed her and punched the down button.

He had hurt her. She was still trying to get herself untangled from the milling dogs when the elevator eased to a stop on the ground floor and the door opened. The dogs burst out onto the truck-loading dock.

"Shit!" Reggie yelled. He jumped forward and began grabbing at the closest dogs and tossing them into the van. Mary Ellen raged inside herself but began to assist Reggie. In a few minutes, all of the dogs were loaded. Dolly helped the injured Clark into the truck.

"Wait here," Reggie ordered and he ran through the door leading to the building's lobby.

It had occurred to him that the building elevators might react to the heat on the sixth floor because of the heat-sensitive call buttons; the kennel attendant would be helplessly trapped if the fire spread too quickly. Reggie found that the elevator had not gone up to the sixth floor, and he pulled the attendant out onto the floor of the lobby.

"Listen, Buster," Reggie said as he glared at the frightened face of the attendant, "if you want to live, just forget what we look like. When The Man asks you, you tell him we wore masks. Right?"

The attendant nodded.

"Just stay put," Reggie ordered, "and keep your mouth shut. Right?"

Another emphatic nod.

Reggie walked over to the fire-alarm box and broke the glass, then ran to the door and headed for the Econovan.

They were well off the university campus before the first fire truck began to move toward the Loventhal Medical Research Center. Reggie stopped the van and stared at the mass of barking, whining dogs in the back. At the outset of the rescue operation, Reggie had told his companions they would place the dogs with other students around the campus or in apartments in Bloomington. But that was before the explosion and fire.

Now he decided to give up trying to find homes for the dogs. "The fuzz will be looking for us. We're dropping them off right now."

Mary Ellen argued vehemently as Reggie drove the Econovan out to a lonely road south of town. She was still protesting violently when he stopped the truck and jumped out. As soon as the rear door was opened, the dogs scampered out into the cold night. Mary Ellen cried hysterically until Reggie dropped her off at her car. Still crying quietly, she headed for her home in Columbus, Indiana, about thirty miles away.

Of the eleven dogs released by Reggie on that February night, five were eventually killed by automobiles, four died from exposure due to their weakened physical conditions after being subjected to the lab experiments, one was killed when he tangled in a fight with a raccoon, and one dog managed to survive.

The dog that made it was a harrier originally obtained from a vivisection supply kennel in Chicago. He had been the offspring of a mating between a breeder's two pedigreed harriers, and had been the runt of an eight-dog litter. He was well-marked with brown, black, and white, but for some frivolous reason he had been named Blackie by the owner's children.

The breeder had been unable to sell him, so he had been sold to the vivisection service for five dollars. The Loventhal Research Center had bought him at the customary rate of three dollars a pound, which amounted to nearly forty dollars for his puppy size. By the time of his release by the students, he was a year old. He had grown to a full adult weight of thirty-five pounds. He was also adorned with six Teflon clamps embedded in various places around his body, but fortunately the experimental sutures did not encumber him, nor did they develop immediate infections. Aside from the pain of experimentation and the trauma of the dognapping, Blackie was in fairly good shape. His body was sound and his senses alert. The night he had been dropped off, he delved into a garbage can, found a few hamburger scraps, and bedded down in a gully. A natural hunter, Blackie survived by catching small game such as squirrels and rabbits in the wooded area of southern Indiana. He supplemented his meat diet with grass, freshly fallen leaves, and occasionally insects, and he roamed freely.

Because man had subjected Blackie to painful experimentation, it was natural for him to shy away from farmhouses and towns. And, because of his breeding lines, which emphasized pack hunting, it was normal for him to seek out other dogs to run with. He nearly teamed up with a couple of pet shepherds, who were feasting on the carrion of a calf that had strayed into the woods and died from exposure. Blackie made the carcass a center for his activities for three days, but when the meat was gone, Blackie left the shepherds and wandered in search of new feeding areas. Eight weeks after his release by Mary Ellen and her accomplices, Blackie met another dog who accepted him and the pair teamed up. By this time, Blackie had begun his transition to a feral state. He was evolving from a formerly domesticated animal to one capable of subsisting without support from man.

Blackie's new companion, a pure-bred Doberman, was by nature and temperament a leader. Blackie followed him through south-central Indiana until spring.

Chapter Three

Mel Ryan eased over onto the shoulder of the road and pulled to a stop just short of the covered bridge leading toward the entrance to Brown County State Park. He switched off the ignition of his car and relaxed for a few moments.

This is going to be better, he promised himself; this job is going to last. Not like Vail.

Mel often woke at night reliving that day when he had encountered that pack of dogs killing the elk. On that day, he had loaded the body of the shepherd in the truck and driven back through the forest toward Vail. When he radioed in his report of the attack and shooting, he had felt that would be all there was to the matter. He had been wrong.

After stopping off at a doctor's for treatment of his wounds, then going to the Red Lion Condominium and chasing off the motley bunch of mutts who were attracted by the bitch in heat, he had driven on back to the Eagle/Vail Animal Control Center. He dropped the dead dog off at the vet's for testing to rule out rabies, then headed for his boss's office to report.

Before he could say anything, his boss motioned toward the private office. The two men walked inside in silence, but as soon as the door was closed, Mel's boss exploded. "You really busted open a can of worms this morning, Ryan. I'm catching shit from three different directions!"

The explanation was short and bitter. When Mel had radioed in the report of the dog pack ravaging the elk, he had also sent in the dog-tag number of the Saint Bernard. A check showed that the dog had belonged to one of the major shareholders in Vail Associates, developers of Colorado's biggest ski resort.

"Now why the shit did you have to shoot *that* particular dog?"

And that wasn't all. Within an hour of the shooting, a U.S. congressman from Iowa came across the carcasses of the Saint Bernard and the elk. The legislator had been cross-country skiing with his secretary/mistress, and the timid little lady had nearly gone into shock when they spotted the carnage.

"You know them Washington types been flocking out here since Ford started coming. Shit! The congressman has been on my ass, and he's got the U.S. Forest Service onto my ass, and the dog's owner wants somebody's head. Ain't you got no sense?"

Mel had fought for control. "Aren't you overlooking a little thing like my responsibility? Our job calls for shooting dogs that are killing wildlife. That's the law."

"Listen, smart-ass," his boss had retorted, "don't go throwing the laws up to me; I helped write them."

Mel could feel a great lump of anger and frustration building up inside him. He tried to explain. "Look, I never as much as used a dog stick before I came out here, and I've come up against some pretty miserable mutts in my life. But these dogs could have killed me. Christ, the dog problem out here in Colorado is awful. Like your big honcho from Vail Associates. Why the hell didn't he keep his dog under control?"

"Maybe he was too busy tending to his job, like you should have been. And it isn't your job to go shooting a major shareholder's dog or making a congressman's cookie barf up her breakfast. We're supposed to *control* dogs, not kill them."

Angrily, Mel yelled, "Then change the goddamned law! 'An animal seen attacking wildlife is to be shot.' That's what it says. What did I do wrong? And if I did what I was supposed to do, what do you want from me?"

His boss shook his head. "It ain't what I want, Ryan. It's what is. You're through."

"You've got to be kidding. I was only doing my job. I didn't want to shoot those damned dogs, but I had to. You're the one who's been yapping about the packs of dogs hitting cross-country skiers. Suppose that pack had hit the congressman and his broad? How about that?"

The argument was strong, but Mel could see that his boss was not affected. He realized that a decision had already been made.

"I'm sorry, Ryan. You said you could handle the job when you came here; you just proved you were wrong. Killing dogs isn't what we're in business for. I'll arrange for your pay."

And that had been that. He'd come home to Indiana to rest and sort things out. When he'd explained the Vail business to his father, he'd been surprised and a little angry at his father's suggestion that maybe his actions had been extreme. He'd refused to discuss the incident further.

It hadn't been easy to find a job that would allow him

to use his skills with animals without dealing directly with dogs. It was only by chance that he had heard about the opening at Brown County State Park, and he was grateful for the opportunity to start fresh. But he'd had to explain the problem he had experienced at Vail.

That interview had taken place in the park superintendent's office a few weeks before.

"Hell," Superintendent Oscar Korteen had said, "we ain't got anything to do with Forest Service people, and that bunch at Vail sure can't say anything to us. Aside from that mess in Colorado, you've got a great background to work for us."

"Thank you," Mel said. "But I hope you understand that that 'mess' in Colorado was no fault of mine. I did nothing wrong."

Oscar Korteen eased himself back into his swivel chair and gave himself time to think by lighting up a cigarette. He liked the looks of the Ryan boy—neat, well-groomed, mild-mannered, and polite. But this last comment might be a subtle indication of a personality flaw. Why did Ryan insist on making a point of his innocence in the Colorado shooting matter? The way the interview had gone, it must have been apparent that the job was available; so why did he have to make such a fuss about the issue? As far as Oscar was concerned, that aspect of Ryan's employment history was unimportant; the kid had gotten fired for reasons other than doing a bad job.

Oscar said, "I've got an opening as my assistant. If I offered that to you, would that convince you I feel you did nothing wrong?"

Mel reflected on the statement. Korteen was probing for something, but he couldn't figure out what. He didn't want anyone holding the Vail incident over his head. Mel hated not working in nature in one way or another. Even more, he hated not working. This was a great chance for him, but he would not take it except on his own merits—and that included accepting the "mess in Colorado" as outside of his control.

Cautiously, Mel said, "I applied for a wildlife officer's job, Mr. Korteen. I don't know if I'm qualified for anything else."

"Look, boy," Oscar replied. "I know when I see a man qualified for a job. Each year we get a few more thousand people coming into this park. The past few years we been getting other problems, with littering and vandalism and even pot parties. So my job is getting tougher, and I have less time for the problems of the forest and the water supply and all of the wildlife control. I think you are qualified, and you could take some of those jobs off of my mind."

So, now, Mel was well into his new job. It was great to be back at work, to feel twigs crack under his feet and hear squirrels yapping noisily as he walked the trails. He put the car into gear and drove through the covered bridge that was one of the attractions of the park. The bridge was constructed of hand-hewn timbers in the manner of the typical structures throughout the park. He eased his car to a halt and smiled at the elderly woman who acted as ticket-seller for the north gate. The Indiana Park Service used the admission fees to help pay for operation and maintenance costs since Indiana is one of the few states who use no tax monies to operate their parks.

Mel drove in, past the road to the swimming pool, and he noticed that some people were already in the park and setting up on the archery range. He skipped his routine stop at the Abe Martin Lodge for a morning cup of coffee; he was anxious to get to the park office because there would be much to do on this day. Oscar had been called to an extraordinary meeting at the Division of State Parks in Indianapolis, and Mel was to function as acting superintendent in Oscar's absence.

The two men had hit it off well since Mel had been hired, and in the short time Mel had worked at Brown County State Park he had initiated the cutting of two new trails, which would not only provide greater access for visitors but would also improve water drainage and cut down on erosion.

"That's the kind of thing we need," Oscar had said when the proposal had been made. "We really have to help Nature out for a while."

Mel had smiled at the comment. The park had been opened in 1929, two years after Oscar had been born and twenty years before Mel had arrived in the world. Surely forty-seven years were enough time for Nature to adapt to the incursions of man. And with over seventeen thousand acres to absorb those incursions, the park should last a while longer. The twenty-seven miles of auto roads and eight miles of hiking trails left much land relatively untouched.

The two men had also agreed on two other major projects suggested by Mel. He wanted to repair and reorganize the Large Mammal Exhibit, which housed deer and buffalo. The facility had fallen into bad shape from neglect. He also wanted to try and shape up the Saddle Barn operation, which had become the private domain of the surly group of employees running the show.

Oscar admitted that he had been lax, and he accepted the criticism without argument. Mel tackled the problems happily. The Large Mammal Exhibit was easy because that merely involved rebuilding some of the pens. The Saddle Barn was more difficult because personalities were involved, but despite minor abrasions to some employees' personalities, it seemed that he would be successful in making some changes.

Now, with Oscar away, all problems fell to Mel. And with only a week to go until Easter, when the busy season traditionally began, there would be much to do. Within days families would begin to arrive to see the full bloom of spring, the tulip poplar and yellowwood would be in blossom, and the newly born animals would begin to roam. The weather promised to be good, so the crowds would build quickly, and the horse trails and nature tours would be crowded.

But as he walked into the superintendent's office that Tuesday morning, the first thing that hit him was a message from the sheriff of Brown County. He had called to ask Mel to attend a meeting in his office that morning.

Even more disturbing to Mel was the subject of the meeting. According to the phone message, they were to discuss the problem of lost dogs.

Mel had met the sheriff during his first week on the job, and he felt the lawman was a decent sort of man. But it still gnawed at Mel whether the sheriff knew about Mel's problems out in Colorado. It was not too likely that Oscar Korteen had mentioned the past to the sheriff. But why did he call on that particular day, when Oscar was away at a meeting? The question did not leave Mel's thoughts as he drank his morning cup of coffee, studied progress reports, and busied himself with administrative assignments. Finally, when he could procrastinate no longer, he left the superintendent's office and headed for town.

Chapter Four

AHA AMERICAN HUMANE ASSOCIATION

ANIMAL CONTROL STUDY

Computer Control Center, Denver, Colorado

EXHIBIT NUMBER 37-14-3A

DOG PARTICULARS

BREED Mixed Lab

COLOR Black

WEIGHT 45 HEIGHT 20

NAME (IF KNOWN) Beau (Roberts)

AGE (APPROX.) 3 years

ANIMAL CONTROL REGISTRATION

Chicago (Cook) CITY (COUNTY) / ILL. STATE / 20027 NUMBER

TYPE OF LOSS Escape: Panic

The Noel Roberts family from Chicago had no idea they were about to become part of an enormous national statistic; they were merely trying to begin a vacation that they hoped they could enjoy. But, with about thirty-five million dogs in the United States, it is predictable that some of them are going to do something to mess up the best-planned holiday.

For the Roberts' three little girls, the trip was a chance to play hookey from school with their parents' permission. For Noel and Sue Roberts, the parents, the trip was a romantic quest for memories; they had gone to Mammoth Cave National Park in Kentucky on their honeymoon exactly twelve years before.

Of course, back when they had first been married, the elegant, easy-to-travel Interstate Highway 65 had not been constructed, so driving from Chicago to Kentucky had been more difficult. This time Noel's pleasure was enhanced by the fall scenery that lined the broad highway ahead.

Earlier in the day, a few hours after they had left Chicago, there had been a tumultuous uproar when daughter Paddy was accused by daughters Lucy and Claire of drinking an extra Coke. Neither Noel nor Sue had been maintaining any type of inventory control or supervised disbursement program, so there was no way to establish the truth. Paddy, of course, insisted on her innocence, though there was circumstantial evidence in the form of an odd number of remaining cans. There was also the possibility of self-incrimination in Paddy's urgent plea, "I've *got* to go to the *toilet!*" But Noel, a lawyer by profession, acted with Solomonic wisdom and replaced the missing soft drink with one purchased when they stopped to replenish the gas tank.

They were riding in their huge Chevvy station wagon with the backseat folded down to give the girls plenty of room to loll around, play games, and annoy their dog, Beau. There had been discussions about putting the dog into a kennel for the week they would be away, but Beau was so much a part of the family that the idea had been summarily dismissed. He was a mixed Labrador retriever who had received every Lab characteristic his pure-bred father possessed except size. His mother had been a small mongrel, so Beau was only about twenty inches tall rather than the ideal twenty-four inches, and he weighed only about forty-five pounds instead of the ideal seventy. But the Roberts family did not really care about his physical dimensions or his lack of pedigree; he was an excellent watchdog, he loved the children, and he was insane over automobile trips.

Beau was incredibly patient with the little girls. While he refused to sleep in the car—or even to lie down while it was moving—he constantly moved out of the girls' way when they tossed their feet, squirmed around, or demanded to sit right where he was standing. It was as if he was so absolutely grateful for being allowed to come on the trip that he would allow himself to be subjected to virtually any discomfort.

"Beau will go crazy christening all of those trees," Sue said to Noel. Beau had a mania for defining his boundaries with his urine, and he seemed to have resource capabilities that measured in gallons, because he was constantly busy whenever they stopped near a wooded area.

Sue caught a glimpse of bright orange far ahead and said, "I think there's an 'H-J' ahead."

Her attempted code for the restaurant went for naught; the three girls exploded with excitement, and it was ordained that they would have to stop for lunch there.

"Okay, okay," Noel chided cheerfully, "we'll stop."

Squeals of glee greeted the surrender; the girls cheered and Beau wagged a furious tail.

The rear-compartment passengers all crowded against the back of the front seat, the girls chattering loudly and Beau rubbing his cold, wet nose against the back of Noel's neck.

Sternly, Noel ordered, "Now, stop the noise. And, Beau, stop soaking my neck." Unaffected, the girls continued what they were doing, as did Beau, who had heard the words but was confident that his master surely did not mean what he was saying.

Noel began to slow the car as he approached the exit that would take them to the Howard Johnson's. He drove down around the ramp, waited at the stop sign for traffic to clear, then turned left to approach the restaurant's parking lot.

As he made his final turn onto the asphalt driveway, a dog came running at the car. Noel had not seen where the dog had come from; it could have been in the deep grass of a drainage ditch, or it could have come from behind one of the cars parked in the gas station. There was no danger because the dog was well clear of the front of the car; however, it ran alongside and then cut in back, very near the rear fender.

Beau went crazy.

It was not the first time he had barked at another dog, and it was routine for Beau to bark when any animal or person came toward the car when the children were inside. But, this time, Beau's bark carried a different timbre, something more violent.

"Beau! Stop it!" Noel shouted just as the other dog slammed into the rear window of the station wagon.

Noel assumed that Beau could not hear the order over his own noise, so he repeated it. Beau stopped barking but, through the rear-view mirror, Noel could see the hair on the back of the dog's neck standing up, hackles raised more markedly than ever before.

"Easy, boy," Noel said softly, and he saw Beau's tail give a few wags. But the hackles remained raised.

Noel pulled the car into the parking area beside the restaurant. He opened the door, and was suddenly deluged with petitions from his daughters. Through the confused trio of voices, he gathered that they each wanted the right to take Beau on his lead so he could relieve himself.

He leaned back into the car and patiently explained that it was a totally strange place, and that maybe this time Daddy should walk the dog.

He was facing the girls and smiling at their arguments, so he really should have seen what was happening. But he didn't.

The other dog had done something—no one ever knew what—but something snapped in Beau's self-control and he made a leap for the front seat. He literally flew between Paddy and Claire before he landed beside Sue, and in almost the same motion he leaped over Noel and through the open door. Noel grabbed for Beau's collar, but the dog had moved with such speed that his hand missed by a foot. Beau was out of the car and running.

Noel yelled, "Beau! Come!" The command had always been effective, but it did not work this time. Beau sped across the parking lot after the other dog.

The next thing that happened was one of those quirks of timing that can never be anticipated.

The other dog had made a rapid escape and, almost as if he realized the chase was futile, Beau slackened his pace. Just then, from off the main road, a huge maroon truck with black trim and bundles of laundry piled on the top turned in to make a delivery to the Howard Johnson's. The driver saw Beau in plenty of time to stop easily, but he was tired and annoyed with his job and slammed on the truck's brakes with slightly more pressure than was really necessary. Beau froze at the squeal of tires and his head whipped in the direction of the truck. At that same instant, the driver smacked his hand down on the horn and, from the force of the fast stop, a bundle of dirty sheets rolled forward, dropped onto the truck's hood, and popped open like a parachute.

The combination of the tires screeching, the horn blaring, and the exploding cloud of dirty sheets fused together in Beau's hearing and vision. He panicked.

One time, as he saw the sheets fluttering toward him, he did hear his name being called. He tried very hard to gather that friendly voice of his master into his mind, but the other sounds were overpowering. Beau ran.

He headed away from Noel and out onto the street. There were no cars coming, so he had a clear field. Noel cursed his cigarette-smoking as his breath faltered, but still he kept after his dog. He thought of running back to get the car, but he felt certain Beau would stop at any moment and he would catch up.

Beau, driven by some mysterious force, was not about to stop. He cut off to the other side of the road and headed up the embankment leading to the Interstate Highway. In a couple of seconds he was out of sight over the lip of the embankment. Noel stopped running and stood, panting hard, trying to decide what to do. From up on the highway, a high-pitched, piercing noise of tires made his decision for him: A car had slammed on its brakes; Beau might have been hit.

Noel ran back to his own car as fast as he could, told Sue to take the children into the restaurant, and sped out of the parking lot toward the entrance ramp to the Interstate Highway. In less than two minutes he had reached the area where the brakes had squealed. There was no sign of Beau, but a woman was standing beside her car, which had slid sideways across the highway and come to rest on the rock-covered verge.

Noel pulled up near the woman and stopped. "Are you okay?"

The woman was visibly shaken, but she managed to answer. "Yes. . . . yes. . . . thank you," she said. Then she added, "Some damned dog just—"

Noel interrupted her. "Where'd he go?" he demanded.

The woman whirled toward him, her face angry. "Was that your dog? You ought to be arrested."

"I'm sorry," Noel said. "But where did the dog go?"

The woman glared, but finally pointed a finger across the highway.

Noel dropped the car into "Drive" and swerved across the grass median. He sped back along the side of the road, looking for his pet.

After searching nervously for several minutes, Noel went back to the restaurant. He told Sue to feed the girls while he looked again for the dog.

He was to spend the next two hours searching for Beau. Finally he had to admit the obvious: The dog was gone. He gave up reluctantly.

He spent another hour with the local police, reporting the loss and reaching out for some encouragement. The Roberts were told to check back by telephone, but little hope was offered: "There's a lot of dogs out there," the desk officer had observed. "No telling if we'll find yours."

The Roberts family sadly continued their vacation, which no longer seemed to be what they had planned. Beau's departure had left a great void.

The sense of loss was reciprocal; Beau found no pleasure in being on his own.

He had heard his master's frantic voice back in the restaurant parking lot, but there were some commands that Beau simply had to ignore. Anything that interfered with protecting the children could not be obeyed.

The screeching and flying white forms had thrown him into a blind panic, and he had simply run—into a strange countryside with no landmarks. He had stopped at last, and had wandered in confusion until weariness overcame him and he slept. When he awoke the next morning after his exhausting, panic-driven excape, his first sensation of loneliness was instantly replaced with pangs of hunger. He was used to pieces of toast, chunks of bacon, and remains of breakfast eggs as the Roberts children competed to give Beau his morning treats.

After less than five minutes of orientation, Beau headed across a nearby field on a compass bearing of three hundred forty degrees—a straight-line course for the Roberts' home in Chicago. The line of travel was such that it kept him well clear of automobile hazards, yet it passed by enough farmhouses and small towns to provide enough garbage for his food needs. It never occurred to him to forage among the available wildlife; predation had not yet surfaced in his personality.

Just south of Seymour, Indiana, he happened across another dog, a young beagle bitch, and they roamed together through the winter.

Eventually they found their way to Brown County State Park.

Chapter Five

Mary Ellen woke and was angry with herself. She had tried to afford herself the luxury of sleeping late, but her reward had been a nightmare. Ever since she had taken part in freeing the dogs from the vivisection lab, she had been having the dream: As the released dogs ran away, their test wounds began to open. Somehow she was running with the dogs, and she could see the Teflon clamps come out, the burn wounds bursting, the splinted limbs . . . But that was where she always awakened, covered with sweat. She tried to urge the image out of her mind by jumping out of bed and rushing in to take her morning shower.

Mary Ellen was the only child of Ken and Shirley Crawford of Columbus, Indiana, about thirty miles east of Bloomington. She commuted three days a week to the University of Indiana, where she was working on her Bachelor of Arts degree in History. Her father, Ken Crawford, was a high-level marketing and sales executive with the town's major industry, the Cummins Engine Company, manufacturer of heavy construction equipment. He traveled extensively, and much of his time at home was spent at the Otter Creek Golf Course, which had been a pet project of the Cummins Engine Company. As a matter of fact, it was often said that the whole of Columbus, Indiana, was a pet project of the Cummins Engine Company.

With her shower finished, Mary Ellen jumped into her favorite Levis, pulled on a sweater, and headed out to round up some breakfast.

Much to her surprise, her father was in the breakfast room, having coffee and reading the editorial page of Columbus's daily newspaper, *The Republic.*

He shook his head at the paper and said, "Balls!"

"Daddy!" she said. "Such language!"

Startled, Ken Crawford peered over the half-lens of his reading glasses quizzically. "What in the world pried you out of bed at this early hour, Mary Ellen?" he asked.

She pecked a quick kiss on his forehead and went to the sideboard to pour herself a cup of coffee.

"Come on, Dad," she said. "I'm not Mom." Since Mary Ellen had been old enough to open her own box of breakfast cereal in the morning, Shirley Crawford had not deemed it necessary to make her appearance before noon, at which time she was either scurrying toward a golf date at Otter Creek or a shopping trip some fifty miles north in Indianapolis. "I, sir," Mary Ellen reported as she poured, "shall spend the first day of my holiday in the worthwhile activity of horseback riding."

She sat down across the table from her father. "Do all men always talk like that? I mean, when they are not around women?"

Ken looked puzzled.

She explained: "I mean, do men always use words like 'balls,' and 'screw,' and 'fu—"

"Hey!" Ken held his hand up. "Where'd you pick up such language?"

Mary Ellen laughed. "I'm in *college,* Daddy." She looked under the table. "Where's Duke?"

She had arrived home after eleven the previous evening and she had noticed then that Duke, her boxer, was not in the house. And now he wasn't at his usual post, begging table scraps.

Ken shrugged. "Oh, I guess he's out sowing some wild oats. We haven't seen him for two days."

"Two days! He never did that before!"

Duke had been given to Mary Ellen seven years before when she had begged and begged for a puppy. The pedigreed boxer had arrived on her birthday, but after the initial excitement, the dog had become Ken's pride and joy. Mary Ellen had not really minded her father usurping her place in Duke's life, especially when the cuddly little puppy reached about half his adult size and managed to maul two neighborhood mutts. It was not until Ken had interjected his strong hand that Duke took on any semblance of a well-mannered animal.

Ken looked at his watch and stood up. "I guess he has a girl friend in town or something. I've gotta run now." He paused, then asked, "Do you want to come to the club?"

Mary Ellen smiled and shook her head. "I'm going to take Arab over to Brown County and see how the trails are."

"Okay," Ken said. "Maybe we can all go out tonight. We'll see what your mother says." He paused again at the door. "Do you want me to put the box on your car?"

Mary Ellen smiled. "I can do it, Daddy. You go on. It's not hard."

Ken knew that hooking up the horse trailer to her Mustang was no big chore, and he was glad she had refused his offer. "Have fun," he said as he left.

Mary Ellen finished her second cup of coffee and decided the morning was too beautiful to waste on breakfast. She went back to her bedroom and got a light windbreaker from her closet. She was slipping it on when she spotted Duke out by the tennis court in back of the three-car garage.

"Aren't you one bad boy," she whispered. She smiled and gave a slight giggle as she noticed that the heavily muscled boxer was leaning wearily against the far net post across the tennis court. Poor old Duke would need some nourishment, and she instructed herself to feed him after she hooked up the horse trailer.

She went out to the garage and rolled up the noisy door as quietly as possible, hoping her mother wouldn't wake. She had already driven her red Mustang out and backed it up to the trailer when Duke finally made his appearance. He was a handsome boxer with a coat the color of mahogany, an excellently deep brisket, and just enough white face hair to accent his black mask. His bearing was generally noble—forelegs straight front, hind quarters moderately angulated, muzzle raised slightly—but today he merely plopped ungracefully down on a soft patch of cool grass and offered only a slight whimper.

Mary Ellen glanced in his direction and grinned at his lazy pose. "So you're a victim of your own insatiable debauchery. Right?"

His docked tail flipped weakly.

She rose, placed her hands on her hips in the joking manner of scolding, and looked sternly at the dog.

It was because he had such a fine black mask that it was difficult to see the blood. When she finally spotted it, she ran to where he was lying. The blood was on his right cheek, over his muzzle and on his skull near the ear. Mary Ellen guessed the wound producing the blood was at the base of the ear; Duke had been injured there before when he had tangled with dogs in the neighborhood. There was no flow of blood, so the cut had apparently coagulated. She decided to clean him up in order to see if the injury was bad enough to require a trip to the vet's office.

"Sit!" she ordered.

Duke obliged by pulling himself up into a sitting position. She walked over to her horse trailer and pulled out an old bath towel she used for wiping down her horse, wetted the cloth at the faucet on the side of the garage, and went to kneel beside the dog. She folded her fingers back and eased her hand to his muzzle as she spoke quietly. "Good boy," she said. "I'm going to see what's happened to his poor ear. Easy, now."

Duke looked at her with expressive brown eyes and let her explore.

She could see no gash or puncture. Gently she touched the wet cloth to his muzzle and softened the caked blood. When it was wiped away she could see that he had no face wounds. She cleaned his entire head in search of the source of blood. There was no wound.

"Well, now," she said. "What have you been up to? Where the *hell* have you been?"

Duke knew the tone and he slipped back down into a prone position, turning his head slightly so that he had to look sideways into Mary Ellen's eyes.

"You come!" she ordered. The boxer rose listlessly to follow as she headed around to the north side of the house, to the expensive, elaborate, heated kennel with its twenty-foot run that Ken Crawford had built for Duke in the shade of the building.

"Inside, young man!" Mary Ellen said in a stern

voice. He trudged past her without looking up and made directly for his shelter. She went back into the house, filled one of Duke's huge bowls with dried feed, and took it back to the kennel.

Duke did not respond to either Mary Ellen's scolding or the filled bowl because he was already asleep.

Chapter Six

AHA AMERICAN HUMANE ASSOCIATION

ANIMAL CONTROL STUDY

Computer Control Center, Denver, Colorado

EXHIBIT NUMBER 10-40-6B

DOG PARTICULARS

BREED Beagle

COLOR Brown/white

WEIGHT 12 lbs. HEIGHT 13"

NAME (IF KNOWN) Mitzie

AGE (APPROX.) 10 weeks

ANIMAL CONTROL REGISTRATION

Indianapolis, Indiana 400683

CITY (COUNTY) STATE NUMBER

TYPE OF LOSS Release/Convenience

The Country Oven Coffee Shop in the Indianapolis Hilton was busy for that time of day. The Speedway had been hosting a meeting of Indy-500 car owners for the past few days, in anticipation of problems over the new carburetor configurations that would be mandatory for the race next May. Most of the owners who had attended the meetings were up early and heading for outbound flights.

Marty Walker checked his cashmere car coat in the cloakroom and walked into the coffee shop. He looked at the unusual crowd, glanced at his watch, and then looked for June.

Marty's Saturday-morning leisure suit was in contrast with most of the breakfasters' business attire, and it was designed for just that: He had spent three hundred dollars to look casual. His hair was expensively razor-cut and his boots were handmade of calfskin.

He looked across the room and spotted June.

June Thompson's face could have been identified by many others in the room. She did an afternoon talk show five days a week and one early on Saturday morning. Her following was sizable in the Indianapolis area.

Marty arrived at the booth, delivered a perfunctory kiss on her cheek, and motioned to the waitress, pointing at the coffee server. Quickly the waitress filled his cup, replenished June's, and left them alone.

As soon as the waitress was out of earshot, June said, "I'm pissed, Marty. I'm really pissed. Can't you get that bitch to do *any*thing?"

Marty was an attorney, a tough, shrewd operator who specialized in the theatrical clients as well as those steel-hard drivers who orbit around the Speedway. He constantly came up against hard people who would give no quarter, and he usually got his way. But not with June, the one woman he had really loved in his whole life. He shook his head. "I begged her attorney and I even called her, but she just won't take them back."

"Great. I'd love a drink," June said angrily.

"Hey," he cautioned. "You have a show to do this morning."

"I canceled. We're running a tape of an old show."

"But why?"

"Do you know those fuckin' pups of yours demolished my rug while we were out last night?"

"No, I didn't know—"

"Marty! That's a seven thousand dollar rug!"

Marty knew it was a seven thousand dollar Persian rug; he had been with her when she bought it to celebrate his divorce just three weeks before.

He suspected the rug wasn't really the problem. June had an inclination to exaggerate, and her "demolished" more than likely would turn out to be a frayed edge or, at worst, a hole that could be patched. The problem was those three damned puppies, which apart from a sizable alimony, were the last vestigial remains of his marriage. Silently he cursed his ex-wife for allowing her to trick him into taking half of the litter of their beagle. The dog had managed to get pregnant due to her carelessness, and after he compassionately agreed to take three of the six pups, he realized that the woman had diabolically planned the whole thing. Marty's lease on his apartment strictly forbade pets, so naturally the pups had gone to June's place, where, in the course of ten days, they had managed to defecate and urinate in every imaginable place, break several pieces of Dresden, and spread garbage and shredded toilet paper everywhere. And now the latest escapade with the Persian rug. Marty couldn't really be angry at June; her life had been hell since he had foisted the pups on her.

"I'm sorry," he said tenderly, "It's been rough, hasn't it?"

"Oh, Martin," she said sadly. "What are we going to do? I'm going out of my head."

Give Martin Walker a million dollar deal to organize or let him take off after a contract violation and he could not be matched; but put him up against a crying woman compounded with the problem of disposing of three beagle pups and he was beaten before he started.

Back when they first took the pups to June's apartment, they had made dozens of telephone calls to friends and associates, but neither was able to find any takers. June even mentioned the puppies on her television show, but no calls came in response to her offer. There was an interval of a few days when they paused in their search for potential owners of the pups, but that came abruptly to an end when one of the dogs up-chucked on June's foot just as she was taking a soufflé out of the oven. The jarring of the pan as it hit the counter caused the inevitable flop, and that led June to the telephone to call the Humane Society. The call to the animal shelter provided the shocking discovery that the animals would be disposed of if not adopted after a certain period of time. Neither of them could face the thought of condemning the puppies to death. Finally, Marty had invested several hours in telephoning pet shops in Indianapolis, but he soon found out that there was little commercial interest in cross-bred pups.

They were still struggling with the problem in the coffee shop.

"I want to leave, Martin," she said.

He did not argue. He motioned for the check.

Once outside, they walked west along Ohio Avenue, silent, looking into windows, each wrapped up in their mutual problem.

They were passing an antique shop, and suddenly June stopped and said, "Look at that, Martin."

In the window he saw a Sheraton table, a Louis XIV chair, a Chesterfield desk, and a Victorian model of a seventeenth-century sailing ship.

"What?" he asked.

June pointed to the side of the window. "There on the wall."

Then Marty noticed what she was indicating, a trio of matched prints in ornate gilt frames, typical hunting scenes in mid-nineteenth-century England. Horsemen stiffly erect in red hunt outfits, and the ever-present pack of hounds.

He was puzzled. "Okay, so we've got a pack of hounds of our own. And no fox."

June was irked. "Don't you see? That's where our dogs should be. We've been trying to find them homes here in the city, and those dogs are supposed to be out in the country where there are fields and a place to run."

"Okay. How?" he asked.

June stood there for a long time, looking at the prints. "We'll just take them out into the country and find them a home," she said.

She spoke with such confidence that Marty failed to see any problems. He merely asked, "When?"

June turned to him, took his arm, and began walking back toward her apartment. "Right now," she answered. "Today we're going to get our puppies nice new homes."

By the time they reached her apartment, she had devised a plan of action. Marty went to the parking garage around the corner to pick up her car, while she went ahead to get the pups ready for their adventure.

In minutes, Marty stepped out of the elevator on the twenty-third floor. June was standing down at the end of the hall, leaning against the doorjamb, looking into her apartment.

He walked up behind her and asked, "What's up?"

She turned slowly and black streaks of mascara streamed down her cheeks. He looked over her head into the apartment.

The word "snowstorm" leaped to his mind.

The sectional sofa had been arranged in an L shape

around a coffee table and in front of a fireplace. A large, stuffed armchair had completed the grouping. All of the pieces had been covered in an expensive brocade, which now lay all over the room in shreds. Manny and Moe, the two male pups, were both innocently asleep. The other pup, Mitzie, stood across the room wagging her short, curved tail. For all of her ten weeks of life, somewhere along the line she had learned something about humans, and she seemed to know something was wrong.

Quietly, Marty said, "I'll take you over to my place, then come back."

June shook her head. "Let's just get these mutts out of here."

Marty left June at the door and crossed the entrance foyer, shuffling his feet through the small mounds of cotton furniture stuffing. He went into the kitchen and removed the three dog leads from their hook in the mop closet. The puppies bounded happily to him in anticipation of a romp outside the apartment.

As they rode down in the elevator, June said, "What are we going to do if we can't find homes for them?" Marty squeezed her hand reassuringly.

In a few minutes Marty was behind the wheel of June's car and heading toward the I-465 bypass that circles Indianapolis. He had given no thought as to where to go, so he simply followed the three-lane highway for a while, then swung south on I-65. There was not much traffic as they drove past the rolling farm lands of south-central Indiana. After nearly a half an hour, June broke the silence by asking, "Where are we heading?"

"Down to the southern part of the state," Marty replied. "I'll bet we can find a nice place for them there."

After a few minutes, Marty said, "I bet they'll find themselves a nice home where they can romp in the country and enjoy their lives. Just look at this country."

June had been looking; it was quiet, peaceful, and inviting. She could imagine the pups having a great time running over the fields. Food, she felt, would be no problem with all the abundant farmhouses set in the middle of well-maintained farms. Surely shelter could be found in the brightly painted, warm-looking outbuildings.

Marty slowed the car as he saw an exit ahead. "I just hope we don't get caught," he said.

"What the hell does that mean?"

"I mean, it's illegal to simply drop off a dog in Indiana. There's a law that says——"

"Oh, stop it!" she snapped. "There can't be a law that says you can't stop them being put to sleep in some animal shelter. That's what we're doing, Martin. We're keeping them from being killed."

Marty did not argue the point. Instead, he pulled the car to a halt on a lonely section of road.

Marty and June sat there for several minutes. In the backseat, the three dogs romped from the seat to the floor to the back-window ledge. Mitzie made several attempts to climb into the front, her tongue licking furiously at Marty.

"We'd better get the hell out of here before a cop comes along," Marty said.

"Should we get them some food?"

Marty shook his head. "The woods are fine for them. They'll get by."

June nodded. She turned in her seat and spoke to the dogs: "You *will* have a good time; won't you, loves?

You'd like to get out right here, wouldn't you?" The puppies wagged their tails furiously. "See, Martin!" she exclaimed, "they do want to play in the woods. See how happy they are?"

Marty forced a smile. He knew she was anthropomorphizing, and he was having qualms about just dumping the dogs out. But he saw no other solution to the problem. He opened the door and got out. It was a balmy fall morning, so he had no misgivings about the weather, and as he looked around the bucolic setting, he convinced himself that this was a safe place for the dogs to begin looking for new masters.

He opened the back door, and it was all he could do to keep the pups from bounding out of the car. But he had read once that it is dangerous to let a dog roam free with a collar around its neck, because the collar could become entangled in bush branches or fence wire. So, one at a time, he removed the collars from the squirming puppies and then stood out of the way. In seconds they were bounding down the roadway toward the wooded area.

Moe cut sharply to his right and squatted in the manner of puppies, but Manny kept running down the road and was quickly swallowed up in the underbrush. Mitzie started to follow Manny, but then she spotted Moe and joined him.

June purposely stayed in the car and busied herself with the improbable task of organizing her purse. Marty stood there for several minutes watching the dogs. Manny never did come back into view but Moe and Mitzie, having emptied their bladders, were sniffing the strange new smells of this foreign area.

"Let's go, Martin," June pleaded from the car.

"Okay," Marty replied. "Just another minute."

He was watching the two dogs and feeling quite content inside himself. His ex-wife had planned to hurt him, but now her effort was being thwarted; his relationship with June had been attacked, but now it could get back to the happy thing it had been; and finally, the pups were out of the city, and they'd be better off here in the long run.

"Look at them, June," he said quietly. "They love it out here."

Just then Mitzie bounded out of a patch of grass and ran over to where Marty was standing. She ran right up to him and sat down at his feet, her tail wagging energetically.

"Hey, little girl," Marty said, "get on back there with your brothers."

Mitzie did a tight turn completely around and jumped up on his trousers. He reached down and stroked her head. Suddenly the pup leaped into the front seat of the car and moved over beside June.

"Martin, let's go!" June pleaded. He could hear an edge of hysteria creeping into her voice. He reached into the car, picked up Mitzie, and walked her over to the grass on the side of the road. He put her down gently, whispered a few nonsensical, friendly words to her, and pushed her bottom so that she'd go toward Moe, who was exploring what was obviously a pungent tree base. Mitzie squatted down and looked back over her shoulder at Marty He could feel his resolve beginning to collapse so he moved back to the car and got in quickly.

As they drove off, Manny was still out of sight and

Moe was vitally interested in sniffing the trees, but Mitzie still sat there, looking longingly at the car.

By the time Marty and June had driven into the city limits of Indianapolis, Manny had made his way out of the protecting sanctuary of the woods and was run over by a speeding car. It had been quick. His skull had been crushed by the front bumper.

By the time Marty and June had reached the solitude of his private club and ordered drinks, Moe had had his left eye clawed out and his throat chewed open when he accidentally wandered into an opossum's den. His had been a slow death.

By the time Marty and June had changed clothes and begun an afternoon of tennis, Mitzie was facing life alone. In the course of a couple of frantic hours, the spunky little beagle had undergone two traumatic experiences: A human owner had rejected her and an animal predator had killed her brother. The rejection was less frightening than the killing, so she opted for staying around man. Even though she was pelted with rocks by people who occasionally did not understand her interest in their garbage, she still voyaged through areas of reasonably dense population. She wandered at random so it was quite by accident that she ran into her first pack-companion, Beau, who had inadvertently run away from his traveling family. She had been in the woods for only three days when her course converged with Beau's; it was her first direct contact with another animal since she had been dumped out of the car.

Beau and Mitzie were an improbable pair: She was less than three months old and not sexually attractive to another dog; he was hardly a raised-in-the-wild animal or even extraordinarily equipped for survival. But they teamed up on a couple of garbage forays and found satisfaction in being together. If Beau had not happened along, Mitzie, young as she was, probably would have lost her battle for survival in the wild. Unlike pampered house pets, who are looked after by owners and ministered to by trained vets, the dog in the wild has a much shorter life span. Where a pet can live easily for ten years and frequently over fifteen, the average stray dog will survive for six to eight years at most. Combat with other more established competitors in the predatory orders, accident, disease—all reduce the dog's chances of survival to some degree.

But Beau was a good provider, and he led Mitzie into shelters that were dry and not windy. She followed happily, because she had done poorly for the first few days on her own. Beau seemed to have a talent for ferreting out particularly nourishing garbage, and he showed her how to tear open sealed plastic bags. And, when it came to dodging rocks tossed by irate homeowners, Beau seemed to have an extraordinary instinct that allowed him to bound out of range at just the right moment.

Through the winter, Beau learned that he could direct his caretaking talents toward Mitzie just as he had toward the Roberts' little girls. When the snows came and temperatures fell below freezing, he even found a temporary home for them at a farm. Fortunately, Mitzie waited until the thaw came before making her way into the farmer's chicken coop and killing three choice pullets. Beau was warned of the incident by the horrible noises coming from the coop, and as soon as Mitzie came running out, he led her off back into the woods and away.

By early spring they had both learned to supplement their garbage scrounging with leaves and grass and even insects, but they were still far from becoming feral. They gravitated toward Brown County State Park partly because there were scents that indicated family habitation.

Visitors to Brown County State Park during the winter months were generally experienced hikers, but some of them had even less compunction about littering than the normal summer visitors to the park. Tossing a bag of waste food into a snowbank seems to violate the countryside less than tossing it onto a barren roadside. But, when the snow was gone, the trash remained, and Beau found it a source of nourishment for Mitzie and him. They established a reasonable roam area covering only a couple of miles in each direction, and they settled down to enjoy their life.

Chapter Seven

The sheriff's office for Brown County, Indiana, is located in Nashville, a small town devoted mainly to such attractions as antique shops, eating establishments, a couple of artists' galleries, and the John Dillinger Museum. Though it is not certain that Dillinger ever so much as drank a cup of coffee in Nashville, this facility is complete with a waxen image of the notorious gang leader laid out in a coffin—a fitting tribute to a man who is believed to have killed about sixteen men.

Mel Ryan parked the jeep in the alley behind the courthouse and walked around to the entrance of the one-story building that houses the sheriff's office and the small block of cells for the infrequent lawbreakers. He paused in the hallway for a moment, then walked across to the window cut into the wall of the sheriff's office. Greasy Fowler was seated at his desk in the back office, polishing his boots. He waved for Mel to come on in.

Certainly Sheriff Greasy Fowler had never had to deal with the likes of John Dillinger, though he probably would have been happy for the chance to do something so dramatic—especially around election time. There was actually nothing "greasy" about the sheriff; he was a tough, sharp, capable officer who might permit his large stomach to protrude over his gun belt but who still kept Brown County a pretty good place for people to live. Many people contended that Greasy had picked up his nickname in honor of Greasy Creek, which joins Salt Creek just east of Nashville and which has a couple of swimming holes in which the local youngsters like to skinny-dip. That contention is just not so, according to the Brown County *Democrat's Almanack.* That publication, considered authoritative in such matters, holds that Greasy was nicknamed as a logical heir to his father, who had been known as "Spoon."

Mel Ryan walked past the booking desk, through the records section past the radio transceiver that provided two-way communication with the department's five cars, and into Greasy's office. He sat down in the metal chair next to the desk and accepted a cup of coffee brought in by Mrs. Fowler.

"How're things going, Sheriff?" he asked.

"Oh, things are okay generally," Greasy said with a troubled look. "I wish you could've made it for the meeting, though. We really got a problem with those damned dogs."

Greasy was used to trouble during the summer, what with tourists flocking in and hippies smoking pot. He was used to trouble in the winter, too, with snowbound roads and hippies smoking pot. And in the spring he had to cope with flooding rivers and hippies smoking pot. But he was not used to troubles during the spring from farmers complaining about dogs.

It wasn't as if his office didn't have its share of calls concerning lost or found dogs and about dog bites, cat fights, abandoned horses, and other animal problems—it did. But Greasy Fowler never had loudmouth farmers climbing up his back, mainly because the south-central Indiana farmers, like most farmers around the nation, take care of their own dog problems—generally with a shotgun or rifle. There had been, in the past, the odd time when a farmer had called for assistance and a deputy sheriff had been sent out to help, but duty reports on such cases were not urgently demanded from the deputies. It was assumed that some dogs got themselves shot and tossed into a handy drainage culvert, but there was something awfully unpopular with the public about shooting dogs, so the matter was traditionally eased into the background.

Mel said he'd been tied up with park business because Oscar had gone to the state capitol. "What dogs are you talking about, Sheriff?" he asked.

"The ones that killed the calf, and hit the chickens and tore through town the other night. Didn't you hear about it?"

"No, I hadn't heard," Mel said. We've had no problems in the park, though."

"I don't guess you have," Greasy said. "I imagine you'd have called me. What I was thinking was maybe them dogs is livin' in the park."

"I haven't seen any," Mel said, "but it would be a great place for them."

There was a silence for a few moments. Greasy had not anticipated that reaction. A couple of years before there had been some dog trouble in the park, and Oscar Korteen had fought tooth-and-nail with Greasy about ever letting the word get out. It had been an unreasonable and tense time between Oscar and Greasy; unreasonable because no one was pointing an accusing finger at Oscar, and tense because Oscar had become vociferously angry when Greasy had deemed it necessary to shoot two dogs right on the edge of the park. In fact, originally Greasy had thought that Oscar had skipped the meeting simply because the word "dog" had been mentioned. But here was Oscar's assistant offering support.

Greasy asked, "You know how Oscar feels 'bout dogs?"

"Not really. I know we have a strict leash rule in the park. Other than that, no."

Greasy leaned forward. "Well, that leash rule was my doing." He went on to explain what had happened. He ended with, "So I don't think you better go saying that too much in front of Oscar."

Mel was sure now that Oscar had not mentioned anything to the sheriff about his problem in Colorado. In fact, this situation could even have something to do with Oscar's easy acceptance of Mel—they had both had problems with dogs.

"What can we do?" Mel asked.

Greasy considered. "Well, there ain't really much none of us can do right now. We'd just like to find out what's up."

"We've had no incidents in the park. Actually, we're just about getting into full swing. This coming weekend will be the first real influx of people."

Greasy nodded. "That's when the trouble will start. Damned visitors let their dogs go crazy. But this is something else. I gotta feeling we've got us a pack of wild dogs."

Mel's training in dog control over the years caused him to balk at that. From all he had seen, most groups of dogs running together are ordinary household pets who are allowed to run uncontrolled. He said as much to Greasy.

"I know. Usually, the only time we got dog problems is when there's a bitch in heat. Then the mutts group up."

"That's my point," Mel agreed. "There's probably a bitch in heat somewhere around." Mel knew that the estrous odor discharged by a bitch in heat could travel incredible distances on slight breezes; male dogs could track the scent for miles.

"Trouble is," Greasy was saying, "stud dogs looking for a bitch don't go killing chickens."

Mel started to say that many dogs in pursuit of a bitch *do* go marauding for sport, but he was cut off.

"It don't matter, anyway," Greasy said.

"It's spring and we got farmers all over the county with livestock dropping their young. There's calves, piglets, and even a few lambs out in those pastures; I've got some pissed-off farmers, and they ain't gonna be happy if I tell them to wait a few weeks till some bitch gets out of heat." He paused.

Mel waited.

Greasy continued. "The reason I held the meeting was 'cause I want to find out where the dogs might be. The vet said just about what you said: The park would be a great place for a pack of wild dogs."

Mel raised his hand in a protesting gesture. "I only said the park was a good range for uncontrolled dogs. I have some serious doubts about actual dogs packing together. Like I said, most groups of dogs are house pets out on a spree. Besides, we've got a leash law in the park and there are too many people working there; someone would have seen them if they were living in the park."

"Okay," Greasy replied. "That sounds okay to me. I was just repeating what the vet suggested. You got any ideas?"

Mel shook his head. "No, but I'll do some thinking."

Greasy stood up, extended his hand and said, "You do that, Mel. I got enough problems without a bunch of dogs raising hell. Another incident and I'll have farmers running around shooting at every dog they see."

Mel said good-bye and left the sheriff. As he drove the jeep out of town he said, "Damn, damn, *damn*! Dog problems again. It's just what I need—people running around shooting at dogs."

Chapter Eight

AHA AMERICAN HUMANE ASSOCIATION

ANIMAL CONTROL STUDY

Computer Control Center, Denver, Colorado

EXHIBIT NUMBER 23-7-7N

ANIMAL CONTROL REGISTRATION

DOG PARTICULARS

BREED Doberman

COLOR Black

WEIGHT 95 lbs. HEIGHT 29"

NAME (IF KNOWN) Jake

AGE (APPROX.) 2yrs +

Evansville — Indiana — B-719

CITY (COUNTY) — STATE — NUMBER

TYPE OF LOSS Escape / maltreatment

Perc Neilson eased his Corvette into the private parking place at the rear of his Shell Oil station on Fulton Avenue, just off Highway 66 north of Evansville, Indiana. He slipped the gearshift into neutral, set the parking brake, and mashed down on the gas pedal. He eased off as the tachometer reached five thousand rpm, held it there for a moment, then pressed down again until the indicator reached the sixty-four-hundred rpm red line. For about five or six seconds, Perc held it there, listening for some audible indication of a malfunction. There was none. Smiling to himself, he eased off, letting the engine rumble back down to its normal eleven hundred rpm idle speed, then switched off the ignition key. The ritual would have seemed normal at a racetrack, but Perc Neilson was forty-three years old and a respected businessman in the community, and it was five-thirty in the morning. People in the neighborhood had reluctantly grown accustomed to the roar of Perc's Corvette each morning at five-thirty, and to his guard dog's irritating barking at various times throughout the night. There had, quite naturally, been complaints, but Perc was 6 feet 2, weighed two-hundred-ten pounds, and was a member of the Evansville Police Auxiliary, which qualified him to mount a flashing light on the top of his car and to wear a police auxiliary uniform on weekends or busy holidays.

As Perc cut off his engine and climbed out of his 'Vette, he was suddenly aware that something was wrong: The guard dog was not barking.

Perc had had this particular dog for nearly a year, and the daily greeting each morning had become part of Perc's ritual in opening up his gas station. If any neighbor failed to be awakened by the rumble of Perc's Corvette, they surely would stir to the yapping of the dog. It was a totally different sound from that the dog created when he was standing guard, but still no one could really call it friendly. When the dog was on duty, the bark was fierce, aggressive, and nasty-sounding; when Perc arrived, the bark turned into a yapping of anticipation, because the dog was fed its daily meal before Perc opened for business.

The dog's name was Jake, and he was a Doberman pinscher just slightly over two years old. He had been working for Perc for a year. Jake had been commercially bred in Decatur, Illinois, by the Guard Dog Breeding Company, which had been accused of unethical practices because of their accelerated inbreeding program. Jake's own lineage was an intricate spider's web of incestuous mating, beginning with his great-great-grandparents. The great-great-grandfather had been shown and had managed to accumulate a respectable number of points at Midwest competitions, and his mate was AKC registered, strong in appearance, and whelped by a champion. The male had cost the Guard Dog Breeding Company eleven hundred dollars, and the bitch had cost another four hundred and fifty dollars. Their litter produced six pups, two of which were obvious runts and were disposed of at birth. The father was then bred with one of the female pups, producing five live dogs, and one of the males from that litter was bred back with his mother. That mating produced four live dogs, and the original male was selected to impregnate one of the bitches, who produced Jake and his three brothers and two sisters. That made Jake have a great-great-grandfather who was also his father and also his uncle, and a mother who was also both aunt and cousin. Critics would rightly point to the undesirability of this practice, and in fact the AKC challenged the registration of the second-generation dogs. Such inbreeding can lead to "bad" dogs; that is, dogs who may be born with physical or mental handicaps. The offspring can display exaggerated genetic traits such as oversized pads or irregular ears, and unpredictable, dangerous, and even vicious temperaments can evolve from too-close inbreeding practices.

The registration problem was solved by issuing phony papers, which satisfied the buyers. After all, the dogs were not being placed in homes as household pets, so the inbreeding did not matter. Besides, when looking at their balance sheets, it was easy to see that this original breeding had, over a period of fifteen months, generated $13,900 in gross revenue. And the company had five other strains of Dobermans who were equally prolific, which would expand the breeding base to the point where the company's ability to train the dogs properly was strained. This, in turn, led to another rationalization: "Make 'em mean and do it fast."

As Perc Neilson walked past the gas pumps toward his building, he thought to himself, "That Gawddamned dawg has got out on me." It had happened to Perc before; a mutt he had adopted from the Animal Control Center had escaped out an open window in the men's room, and that had led him to buy Jake. That's what bothered him most, because he had paid the Guard Dog Breeding Company $825.00, and that was a hell of a lot of money.

As he turned the corner of the building, it wasn't hard to see what had happened. During the night someone had taken a crowbar and burst open the door lock, then knocked off the heavy padlock.

He moved in carefully, wondering if the dog had gotten out or if it had been killed. He was about four feet from the large glass door when he noticed splotches of blood on the inside of the glass.

"Shit," he said out loud. "Eight hundred 'n twenty-five rotten dollars shot to hell."

His experience as an auxiliary policeman had trained him not to go inside or to touch the door, because there was always the possibility of a novice crook who might leave fingerprints. He walked past the office entrance door and walked to the big garage door at the grease-rack bay. He peered through the glass and felt as if someone had poured cold water down the inside of his spine: Crouched under the tire repair machine was Jake, with a small pool of blood forming about him on the cement floor.

Suddenly Jake lunged forward as if he had been propelled by a strong catapult. *It was weird,* Perc thought to himself in the instant required for it all to happen: Jake was in the air, looking as if he was traveling parallel to the huge lift-up door; then, strangely, the dog seemed to change direction while airborne, and slammed against the glass right in front of Perc's face. Blood splashed on the window where Jake's body had struck it.

As Jake fell to the floor, recovered, and lunged at Perc again, the confused man shouted: "Hey! Stop that, you Gawddamned dawg!"

The words and even the irascible tone of voice should have been familiar to Jake. The man who had sold him the dog had warned him not to make a pet out of Jake. "I wouldn't try to kick the shit out of him," the man had warned Perc. "You're likely to get a foot torn off if you do. But, remember, this animal ain't no friend of yours. He's working for you, so don't be nice to him."

As Jake leapt again and spattered more blood, Perc began to worry that his investment might bleed to death right in front of his eyes if the dog kept jumping at the glass. Perc moved back to the entrance to the office and went to open the door. The passageway from the office into the grease-rack bay had a door, but that was normally kept open at night to allow Jake to guard the whole building. Perc had the front door opened about three inches when Jake charged into the office. This time, instead of ineffectually crashing against the glass of the door, Jake pointed his nose right at the slight opening and drove hard as if to wedge his head through to clear the way for his body. Fortunately the office was floored with a smooth ceramic tile, and Jake's pads slipped. He caromed off the doorjamb and knocked over a wastebasket. The impact and strange noise shocked Jake into inaction for a moment, and Perc was able to slam the door shut. At the same time he saw what was causing the bleeding: a slight gash over the dog's left ear. The burglars must have been ready for Jake, and one of them must have clubbed him with something. But the wound seemed only a minor injury, and shouldn't be bleeding so much. Unfortunately, a common trait developed from consecutive inbreedings is canine hemophilia, which can cause excessive bleeding even from a small cut.

"Food!" Perc yelled in through the door. That shouted command was one thing that had always worked with Jake; that had been part of his training. But this time Jake merely gave another jump, snarling, growling, and baring his teeth. Perc realized that the dog was probably going to do himself more of an injury, so he slipped around the corner and out of sight.

Perc stood there for a few moments, smoking a cigarette and trying to figure out what to do. At first he thought of calling the Guard Dog Company in Decatur, but he had done that about six months before with amazing results. ("The damned dog bit me!" he'd complained. "Well, Perc, if we'd known you wanted a chihuahua we'd've sent you to a fancy pet shop in Chicago." the salesman had said, laughing.) Then he thought of calling the vet who had given Jake his shots. ("Mr. Neilson, you'd better give some serious thought to not keeping Jake. We had to tranquilize him just to give the shots. That'll be sixteen dollars for the shots and twenty-three for the tranquilizer.") Finally he decided to call his friends at the police department. Surely they'd come up with some idea.

The telephone booth was over at the corner of his station property, well out of the line of sight from where he had last seen Jake. When he'd made the call, he eased himself past the telephone booth slightly till he could see inside the garage. Jake had moved back to his defensive position under the tire repair machine. Even though he was nearly sixty feet away, Perc could see the white flare of teeth as Jake snarled again.

It was nearly forty minutes before the police car arrived, because the radio dispatcher had called Sgt. Tom Olsen, head of the Evansville Police Canine Patrols, and he was all the way in the south section of the city when he received the call. In the intervening time Perc's three employees had arrived, and they were having a good laugh at Perc's expense. After about twenty customers had to be turned away from the gas pumps, Perc went and sat on the rear deck of his 'Vette and brooded until Sgt. Olsen arrived.

The blue and white car eased up next to the Corvette and Sgt. Olsen climbed out. "Hey, Perc," the sergeant called. "I hear you got troubles."

Perc walked toward Olsen. "You bet your ass I got troubles. My dog's gone nuts and them niggers knocked over my station!"

Olsen looked both perplexed and perturbed. "They didn't tell me about a robbery," he said.

"Yeah," Perc replied, "but I forgot to tell them when I called."

Olsen went back to his car and radioed the dispatch desk for a burglary car to swing by. He signed off, then went back to Perc. "Now, fill me in," he ordered.

Perc told him the whole story.

"Well," Olsen said, "I'd hate to mess any clues up for the detectives, but they sure as hell can't go in as long as your dog's going bananas. Let's get him out."

"Can I help?" Perc offered, hoping he couldn't.

"Naw," Olsen replied, "I'll just get some things from the car." For the next couple of minutes, Sgt. Olsen busied himself donning a padded coat, putting on a fencing mask, and pulling on thick hockey gloves. Finally he assembled his dog stick, a six-foot-long hollow pole with a loop on one end and a pull cord running to the opposite end.

A small crowd had gathered out by the sidewalks, and just as the two men rounded the corner and headed toward the entrance to the office, another blue and white patrol car pulled into the station. Sgt. Olsen ordered the two patrolmen to keep the crowds back and to tell the detectives to stay out until he had the dog under control. He also asked them to call for the Humane Society pickup ambulance.

As soon as Olsen turned to approach the building, Jake began a repeat performance. Olsen waved Perc back out of sight of the dog, then stood perfectly still for a couple of minutes, studying the problem. It was apparent that the dog had a bad wound, because blood was still showering on the window. It was just as obvious that the dog was in a bad mental state because, even though Olsen made no move toward the building, the dog still continued his vicious attack on the glass. Olsen moved to the office entrance and tried the handle. Despite Perc's warning, Olsen was not quite prepared for the degree of violence with which Jake slammed himself against the glass door. For a moment, the sergeant began to think that maybe he had finally found a dog that might need to be shot.

As Olsen studied the problem, he saw that he would need two hands to work the dog stick and a third hand to handle the door. He would have liked to use one of the patrolmen, but they were having their hands full with the gathering crowd on the sidewalk. Ideally, he wanted another dog handler there, but time was short if they were going to get the dog medical assistance. Reluctantly, he called to Perc to join him.

"Now, look, Perc," Olsen said. "You open the door. Just a crack, mind you. I only need an inch to get the stick inside. Okay?"

"Gotcha."

Eagerly, Perc reached down and took the handle of the door. He eased it inward about an inch and stopped when Olsen ordered, "Hold it. Right there."

Olsen felt he could take advantage of the wide expanse of glass in the door to see what he was doing; it should be an easy job to slip the loop over Jake's head if the dog came in range of the device.

Jake had not stopped his explosive attack on the glass door. Perc felt cold sweat trickle down under his arm as Jake hit the surface just four or five inches from his throat.

That Jake was not acting in blind panic first became

apparent when he dodged around at the end of the dog stick as Olsen made his first pass with the loop, then used his powerful jaws to grab at the stick a full eighteen inches above the loop.

"Holy Jesus!" Olsen shouted as he felt the yanking on the stick. "Hold that Goddamned door, Perc!" Olsen struggled to maintain control of the stick. After a couple of hard twists and pulls, the sergeant freed his instrument and swiftly extracted it from the crack in the door. He looked at the heavy-gauge metal pole and he was shocked. Jake's teeth had dented the aluminum and, amazingly, there was a four- or five-degree bend where the dog had clamped down and tried to pull the dog stick out of Olsen's hands.

"That's one strong dog," Olsen said admiringly.

Perc replied, "Hell, I paid eight and a quarter for him."

"Well," Olsen grunted, "Let's try to save him, but he damn sure doesn't look like he wants any help."

Olsen pulled off the glove from his left hand and held it up. "Open the door again, about three inches," he said. "I'm going to throw this in and see if I can trick him."

The door went ajar and the glove flew inside. Jake descended on it with just as much violence as he had displayed before. His teeth sunk into the leather and his head jerked furiously. Olsen whipped the loop down onto Jake. He flicked the handle slightly to the left and had the stick in place. With his right hand, he gave a gentle pull on the setting cord to tighten the loop.

There was an eerie howl, and Olsen saw right away what had happened: The loop had only gotten partly on the neck; the rest of it was curved up over the skull and was trappd on Jake's cropped ear right where his wound was gushing blood.

Olsen struggled to keep Jake under control. He yelled at Perc, "Open the damned door. Now!"

Perc had to push against the pneumatic closer and consequently had to ease part of his body in to the doorway. As soon as there was enough room, Olsen took a couple of steps inside, moving as carefully as possible so as not to hurt Jake any more than was necessary. The dog weighed over ninety-five pounds, and it was solid muscle, but the pain of the ear wound aggravated by the loop made him nearly weak.

Jake bent his head down and, almost like a playful puppy, tried to use his right front paw to push away this new hurt by his ear. His howl subsided to a whimper as he tried again.

Olsen moved fully into the office and stepped around in back of Jake, holding the stick cautiously. "You stay back," he told Perc.

"I want to back his rear against the desk or the wall so I can slip this loop all the way down onto his neck."

Before he could make another move, Perc said, "Shit, I'll git his ass where it belongs." His big foot landed solidly on Jake's flanks.

The kick and the familiarly rough voice were enough to snatch Jake's attention away from his hurt ear. He twitched his body to the left, jerked his head down and up—and the loop slipped free.

Perc's kicking foot had barely touched the ground when Jake made his leap. He hit Perc in the chest, snarling, and then bolted out of the door. Olsen yelled, "You stupid sonofabitch, you let him get away." But then Perc, obviously in shock, turned full face to Olsen, and he forgot his anger. It wasn't a nice wound, and surely not one you wanted to look at after you had just called someone a sonafabitch. Jake had managed to hit Perc just above the left eye, and his long canine tooth had punctured the cornea and then raked down the cheek. Despite all of the gushing blood, Olsen could see Perc's cheekbone bared by Jake's attack.

Outside the station, Jake loped easily toward the street. Most of the crowd began scattering as he approached, but one unwise man thought he could grab the dog. The attempt earned the man a broken index finger.

Jake made it across Fulton Avenue and ran at full speed along a drainage ditch for a half-mile. Tiring fast, he found a large, cement storm sewer to crawl into. Some ancient survival instinct told him to compress the ear wound, and he spent the night sleeping with his left paw pressed firmly against the injury. By morning the blood had coagulated, and the wound had begun to heal.

A few days later he happened across Blackie, the harrier released from the vivisection lab at Indiana University.

Jake and Blackie developed into an effective team. Blackie's breeding had equipped him to locate prey, and Jake's had made him into an efficient killing tool, more than a match for any animal in the woods.

Jake taught Blackie to kill more than just rabbits. Together they ventured upon a group of strays outside Evansville. The strays were all pets, properly tagged but allowed to run free by their owners, and on that day they had come together out on Morgan Avenue near Oak Hill Road in search of a bitch in heat. Jake and Blackie had gravitated to the odor, too; in fact, not much known to man or dog could have kept them from investigation of the smell. Arriving on the scene later than the other dogs, Blackie hung back as Jake ran, snarling, at full speed into the pack. A black and tan coon dog attempted to defend his place in line, but Jake ran in with his head low and thrust up under the coon dog's belly. With one swipe of his wide-open jaws, Jake slashed the dog's scrotum and sent him howling off minus one testicle. At the same instant a springer charged at Blackie and the ensuing clash left the springer minus an eye and half of his right ear. Blackie had tasted his first canine blood. The remainder of the local courtiers offered various minor attacks and were rewarded with sundry wounds ranging from nipped legs to sliced rib cages. Snarling and growling fiercely, Blackie kept the other dogs at bay as Jake mounted the bitch. Hours later the two dogs moved north out of Evansville, leaving behind a group of damaged suitors and a bitch massively impregnated with Doberman genes.

Within several weeks Blackie was beginning to swing over to a feral state, while Jake's attitude remained attack-oriented, as he had been conditioned during his life as a domesticated guard dog. Efficiently teamed, they probably would have roamed over southern Indiana for their entire lifetime without ever really settling down. Then, in early spring, they picked up the first whiffs of a bitch in heat when Mitzie, comfortably ensconced with Beau in Brown County State Park, began her first season.

Chapter Nine

Mel was troubled during his ride back to the park after the meeting with the sheriff. He had admitted that there could be a dog problem in the park, but there had been no manifestation of dogs so far and that troubled him. He remembered back in Boston, when he had been seriously working in animal control, there had been a lecture by an expert from Alabama. The most striking thing Mel remembered was that truly feral dogs maintain an extremely low profile. During one study, radio transmitters had been attached to collars and put on dogs that were positively identified as feral. The pack was tracked via telemetry, and on one specific occasion, the wild dogs remained hidden in their base area while barking hunting dogs chased a deer about a thousand feet away. Another trait had been confirmed in the study: Wild dogs tend to be nocturnal, and that makes them unlikely to be observed casually. In other words, you would have to go looking to find a pack of feral dogs in the woods.

Instead of going into the north entrance of the park up past the Ramada Inn, Mel drove south on State Road 46 so that he could enter the park at the west gate. He stopped and asked the guard on duty if he had seen any dogs, but there was nothing to report from there. He drove on in past the west lookout tower and took a work road down to the water plant; no one there had anything to report. He cut through the closed access roads around Lake Ogle and came out at the Horseman Campgrounds. Several cars were parked with horse trailers open and empty. Only one person was unloading a horse: a young girl whose expensive horse box was attached to a snappy-looking Mustang.

Mary Ellen had picked up her horse at the private stable near her home in Columbus and had driven over to Brown County State Park to meet some of her friends. The stablemaster had told her that her friends had called to cancel because of various problems. Obviously none had wanted to call her at home and risk the wrath of her mother being awakened before noon.

Mary Ellen had decided the day was too good to waste, so she had come on to the park alone. She was halfway through handling her tack when a tall young man walked up to her and offered to help. They exchanged names, and Mel asked, "What's *his* name?" He slapped the horse's rump lightly.

She grinned. "Sometimes I call him Stupid, but generally he answers better to Arab."

Mel could spot an opening gambit when he heard one. He stroked the bright white mane of the horse.

"He bites," Mary Ellen warned as she adjusted the stirrup. To show he knew horses, Mel continued to fondle Arab's head, scratching the forehead. Mary Ellen moved around to the other side to finish her adjustments.

Mel said, "I've just started working here. How do you like the trails?"

"*Dee*vine," Mary Ellen replied, "I just love it here."

They continued to chat as she completed adjustments to Arab's tack, and as she prepared to mount, Mel made a decision. "I've been meaning to take a better look at the trails," he said. "Could I ride out with you?" He explained that it it would take him a few minutes to check in with his office, then he would saddle a horse and ride back to meet her.

"Sure," she said. "I haven't ridden Arab in simply months; it'll take me a while to get him manageable again, so take your time. I'll be here—unless he decides to gallop off with me."

He smiled and went back to his jeep. As he drove toward the barn where the Park Service horses were stabled, he watched her in the rear-view mirror. Arab was trying his best to give her a hard time, but she was winning the battle of wills. Arab's bucking and rearing had been reduced to a well-disciplined canter by the time Mel returned on a large sorrel mare that was kept for travel where motor vehicles were not satisfactory.

As he rode up to her he cautioned, "I hope Arab doesn't mind a lady horse."

She laughed. "Arab's no gelding but he *has* always seemed a little. . . ." She finished the sentence with a limp-wristed gesture. "It looks like he might be switching, though," she added as Arab inspected the mare.

Before Mel could answer, she nudged Arab's flanks and loosened the reins. The horse leapt forward. By the time Mel was able to swing his mare around, all he could see of Mary Ellen was her long blond hair flowing back as Arab disappeared down a trail heading for Smith Overlook.

The mare was less spirited than Arab, but Mel caught up to Mary Ellen in a few minutes, mainly because she had pulled her horse in to a more gentle walk. The peaceful trail was wide enough for three or four horses abreast and Mel slipped in beside Mary Ellen without talking. Arab skitted obliquely as the mare came up head to head, but Mary Ellen displayed her skills by bringing him quickly under control.

They pulled up for a pause at Smith Overlook, awed by vistas of southern Indiana landscape in bloom. The yellowwoods were exploding with blossoms, and the tulip poplar and hickory filled the air with spring smells.

"Do you ride here often?" he asked.

Mary Ellen had been lulled into daydreams by the warm spring morning and the ride. Mel had to repeat the question.

"Oh, sure," she said. "A couple of times a week during the summer."

"Is there anything you don't like about the park trails?"

She gave him a broad grin. "Should I fill out a questionnaire?" Then she saw his uncomfortable blush and quickly added, "No, I was just kidding. I really do like this park. It used to be bad in the gullies, you know, where the run-off water would wash away the dirt and expose the rocks. That used to be bad for the horses, but they did something and it's much better now."

Mel filed that and decided to see if the erosion problem could be improved further.

"And there is that stupid rule about fallen trees and limbs," Mary Ellen went on. "Everyone thinks that's dumb. To roast hot dogs you have to bring charcoal from home so you might as well stay in your own backyard."

Mel made a mental note to forget that complaint; the park had a rule that nothing should ever be used from the ground to put in the cooking pits. The purpose was to allow the fallen branches and trees to create a humus for future growth. It was a good rule and should stay, but he offered a polite nod to encourage her to go on.

They rode casually for another hour, sometimes discussing the problems of the park, sometimes not talking at all. As they meandered eastward in the park, Mary Ellen showed him a hidden trail that went steeply downward and hooked around toward Greenhorn Creek. She led the way.

It was a sharp descent, and brambles had grown over the space during the winter. As they neared the bottom, Mel said, "You shouldn't use this path, Mary Ellen. Someone could be hurt."

She twisted herself around in the saddle. "We do it all the time," she said. "The other kids and I, you know. It's not bad after this, and it is less crowded on busy days during the summer."

Mel nodded, but made a private decision to have a retaining fence installed at the top of the path; if this route became too well-known, some rider was going to end up with a broken neck or an injured horse.

At the bottom of the path was a small clearing and a crystal pond surrounded by a ledge of grass-covered rock. Without discussion, they pulled up their horses and dismounted. Arab, obviously familiar with the haven, strolled over to the pond and began taking a drink. The mare followed.

Mel perched himself on a rock covered with cool, slightly damp moss. Before sitting down, Mary Ellen pulled off her sweater, revealing the outline of well-formed breasts and a flat tummy underneath her tied shirt. He reluctantly forced his attention back to the clearing, and she sat down on the rock beside him. He could see her reflection in the water, though the light-blue eyes and golden hair were already imprinted in his mind.

They sat quietly for several minutes.

Then, at the edge of the clearing, there was a movement, a small thrashing in the undergrowth. Mel rose and walked across to the spot. Mary Ellen followed. She caught up with him just as he said "Oh, damn!" In a heavy patch of briars was a tiny bunny, captured in a prickly branch, its left foreleg sliced neatly off at the shoulder.

Mel bent down and reached in, scratching his arm on the briars, and managed to grab the bunny by the nape and lift it free. He placed it gently on a pad of grass, and the poor creature made a futile effort to hop. It was then that Mel saw the bunny also had a broken right hind leg.

"Turn around," he ordered Mary Ellen.

She turned and tried to set herself, but still she flinched at the *crack* when Mel snapped the bunny's neck. By the time she turned back, Mel had laid the bunny back on the ground and was bending over, studying the dead animal. He stood and turned back towards the thicket. Mary Ellen followed. "What is it?" she asked.

"There's a warren somewhere nearby; that poor thing couldn't have traveled too far. God knows why it hasn't bled to death before now." He peered into the underbrush, then took her hand as if it was the most natural gesture and led her into the brush. A few yards in they found the rabbit warren: six dead bunnies and their dead mother were strewn around the twig-covered opening of their violated nest.

"Goddamned dogs!" Mel said angrily.

"Dogs?" Mary Ellen asked. "How do you know it was dogs?"

Mel pointed to the top of the warren. "There," he explained, "see the torn-off bushes, the claw marks where they dug into the burrow? And the rabbits—cut at the throat, bowels ripped open, and only very little of the meat eaten. It was dogs."

Mel knelt down and swept aside some leaves from the area near the carnage. Mary Ellen continued looking at the dead rabbits with a fascinated sense of unreality. She thought of her own dog, Duke, and then instantly erased the gruesome picture. The park was nearly fifteen miles from her home; besides, Duke would not do such a ghastly thing. "Are you sure it was dogs? What about coyotes, or foxes, or some other animal? Surely there are other predators. Why did it have to be dogs?"

"First, predators kill to eat; not for kicks. And see here." He indicated a track in the mud near the dead rabbits. "The pad marks are broad; a coyote's would be noticeably oval. And the claw indentations are different, too. Only two claws show with a coyote, while all four show from a dog's claws."

He stood and looked at her, his eyes hard and with none of the tenderness they had shown only a few short minutes before. He moved his head in a slight jerk that suggested they leave.

"Shouldn't we do something?" She asked. "Like bury them?"

"Nature will take care of that," he said. Then he

added, with anger in his voice, "I wish Nature would do something about dogs."

Mary Ellen turned abruptly and pushed her way through the bushes. When they reached the clearing, Mel took a couple of minutes to return the one dead bunny to the area of the warren. By the time he rejoined Mary Ellen, she had mounted her horse. She looked down at him.

"Why do you want to leave them like that? Won't whatever killed them just come back and feed on them?"

Mel sensed the edge in her voice, but he tried to ignore it. "Burying animals in the woods is silly. A carcass fits into the food chain; it would be stupid to interject human feelings into a cycle that Nature has developed. And, as far as the dogs are concerned, don't worry. They won't come back."

"How can you be so sure?"

"I'm not sure. But pet dogs only seem to kill for the fun of it; they'd rather get their nourishment at home, from their bowls, in a warm kitchen."

"Pets! You think *pets* would do that?"

He looked up at her and sighed. "Listen, Mary Ellen, I love dogs as much as anyone can. But you've got to realize that dogs are fantastically well-equipped to attack and kill. We've had them as companions for thousands of years, and yet they still retain the predator genes that make them potential killers."

"You can't love dogs and think of them that way."

He shrugged, "I deal in facts. I've had to." He went to his mare and mounted up.

"I see," she said. "I suppose you must have met some nasty dogs in your job, but pets aren't killers."

He thought of Vail, but said nothing. No sense in confirming her suspicion that he was anti-dog.

They rode in silence back toward the horse barn. About halfway there, Mary Ellen asked, "What about wild dogs? Could it be wild dogs?"

Mel pulled himself out of his own deep contemplating. "It could be, but it was more likely pets. We have a pretty good control over dogs in the park as far as I've seen. No dogs are allowed at the lodge, and only leashed dogs are permitted in the general park area. But the dogs that killed those bunnies—I'd guess there were four or five from the looks of the area—could be nice, lovable pets from around here."

"From where, Mel? Where could they come from? There aren't any houses near here."

"Hell," he said. "Nashville is only a few miles away; I've known dogs that could cover ten to twelve miles in an hour to get at a bitch in heat. We're close enough to Columbus, or even to Bloomington, for them to roam in."

Mel had become distracted by the dead rabbits and the possibility of dogs in the park, and he unsaddled the mare in silence. Their parting was cordial but cool.

Mary Ellen began to dismantle Arab's tack as Mel drove away in the park jeep.

Within the next hour and a half, she dropped off Arab at the private stable, drove home, found her mother gone for the afternoon, and made herself a late lunch. She sat on the patio by the pool to eat her sandwich, but then she heard Duke barking in his kennel and went around the house to let him out. He followed her back to the patio area.

As she finished her sandwich, Duke sat there drooling and making his customary awful begging sounds, which had been encouraged by her father ever since the boxer was a puppy. She tossed him a crust of bread, which made him whine even more.

"No! Bad boy!" she scolded.

He looked properly chastised for a moment, then he begged again.

As she finished her sandwich, Duke realized any hope of further handouts was past. He trotted across the patio and jumped into the pool. After a five-minute swim, he came out by the stairs at the shallow end, waited until he was within sprinkle range of Mary Ellen, and shook off the water. Then he stretched out on the warm patio surface.

With a serious tone, Mary Ellen said, "Duke, would you kill a little, helpless bunny?"

Duke lifted his drowsy eyelids and looked at Mary Ellen. He knew she was saying something, and it did not sound like a rebuke so he permitted his nub of a tail to flip back and forth tentatively. "I didn't think so," Mary Ellen said. "You're a good boy, aren't you?"

"Good boy" was a well-learned collection of sounds, and the tail wiggled more. Then Duke eased his eyes shut with the hope that the session was over and he could go to sleep.

Mary Ellen settled back to enjoy the sun, but every now and then she glanced down at Duke and tried to imagine an infant rabbit being mauled in his jaws.

It was near quitting time when Oscar Korteen arrived back in his office from the meetings in the state capitol at Indianapolis. Mel was waiting for him.

While they prepared cups of coffee, Oscar complained. "Hell, the director was planning to raise the admission to the park to two dollars. Ain't that ridiculous?"

Mel crossed over and sat down on the couch without comment. Oscar moved behind his desk, and began glancing through the mail, but he continued his complaint. "Do you realize that half the kids who come to the park would be cut off? And the bastard told me we'd be seeing no increase in our budget, either. The money has to be used by those idiots in the legislature for welfare appropriations." He snorted. "Well, I told him he can get a new superintendent if he does that. By God, he might have to find a whole new park. People would just stop coming if the admission was raised too high."

In the ensuing pause Mel said, "I rode some of the trails today."

Oscar ignored him. "Well, he finally saw the light. But we're still going to have to increase some fees, like the camp store concession—though God knows how Lilly is going to be able to handle that; she makes little enough. And the charges for the swimming pool and saddle horses and overnight campers will be going up. Damn!"

Another lull. "I found a big problem," Mel said at last.

"*You've* got a problem!" Oscar nearly yelled. "Don't you care about this park? I had a hell of a time today—"

"I *do* care," Mel insisted. "That's what I'm trying to tell you. I do care, and we've got a problem. Here, in the park."

Oscar frowned, then gave a sigh. "Okay," he said. "What'd you find?"

As Mel went through his story, Oscar's face moved

from interest to concern to something like anger. Clearly, he did not like what he was hearing.

When Mel had finished, Oscar rose from behind his desk and went to the coffeepot on top of the pine book shelves. He did not rush as he held the spout over his cup; he was thinking, and he looked around the room he knew so well. The building was an authentic log cabin, though the original mud chinking had been replaced with modern plaster. His desk had been constructed from split logs, the chairs were fashioned from rough wood, and there were pictures on the walls that dated back to 1929, when the park had been opened. Oscar liked the office and wanted to stay there. It had come to feel like home to him.

He went back to his desk.

After several moments, he said, "We've never had any major dog problems, Mel. Do you think this could be just an isolated incident?"

"Hell, Oscar, of course, it's only an incident. I'm telling you about it because the sheriff seems to think there is a problem in the area, but I don't know how we can call one rabbit kill anything more than an incident."

Oscar asked, "Could those dogs have just been passing through?"

"Sure."

"Then there's no reason for you to get concerned, is there?"

Mel was beginning to hear a tone in Oscar's voice he didn't like. "Look," he said. "Let's put this in some kind of proper perspective. Some rabbits killed is only one small thing, but it is a *fact* that the killing was done by dogs. The thing I'd like to do now is go out and take a hard look around. I'd like to see if the dogs are using the park as a base. I could probably find out in a day or so."

Oscar leaned back and studied Mel. "You been talking with the sheriff a lot?"

"Only at his office this morning."

"Well," Oscar went on. "you might have found out that I don't have much use for the sheriff and his attitude toward dogs. You know he killed two of them up by Salt Creek, right near the north gate?"

"He told me."

"He did, huh? Well, I'm telling you that I don't like the whole thing. There's no need for you to go wasting your time looking for a pack of dogs. I told Greasy Fowler that two years go, and—"

"But he found some dogs to shoot," Mel interrupted.

Oscar jumped up and paced across the room. "Greasy Fowler found two simple mutts who weren't hurting anybody. That guy is nuts about shooting, and he used my park for a target range. I got into a bunch of hell about that, and I *don't* plan to let it happen again."

"It would only take me a couple—"

"No! You've got a lot of work to look after. Easter is less than a week away and we'll be flooded with people. There's plenty to be done."

The two men looked at each other. Mel admitted to himself that there was a great deal of work to be done in the park, and it was his job to supervise it. But the dogs were somewhere, and there could be bad problems if they were in the park, especially with the spring rush about to begin.

"Look, boy," Oscar said in a softer tone. "I know what you went through out in Colorado, and I know that kind of experience can make a man a little jumpy. But you don't have any real evidence that we have a dog problem here, so let's just let it alone. Okay?"

"What happened in Vail has nothing to do with this," Mel said coldly. But as he sat there, he began wondering if, in fact, he was not overreacting to the situation. Ever since the incident in Colorado, he realized that his attitudes toward dogs seemed to be changing.

He stood up and forced a smile. "Maybe you're right, Oscar. And you're sure right about the work I have to do. I'd better get to it."

But, as he walked away from Oscar's office, he knew it was not over. The dogs were out there, and they'd move again.

Chapter Ten

AHA AMERICAN HUMANE ASSOCIATION

ANIMAL CONTROL STUDY

Computer Control Center, Denver, Colorado

EXHIBIT NUMBER 91-05-P-7

DOG PARTICULARS

BREED American Pointer

COLOR

WEIGHT 55 lbs. HEIGHT 26"

NAME (IF KNOWN) Clum

AGE (APPROX.) 18 months

ANIMAL CONTROL REGISTRATION

Paducah — CITY (COUNTY)
Kentucky — STATE
- none - — NUMBER

TYPE OF LOSS Abandonment

Jace Jones was a typical Kentucky farmer—hard-working, tough, and shrewd. He had been on the family farm in Ballard County for all of his life, and he knew how to make his land pay. He grew both corn and hay, which he sold for feed to cattle growers at the market in Paducah, but his money crop was tobacco. Jace Jones knew how to grow his burley tobacco, and because he knew, he made money. Jace used his money for his farm, children, and dogs; he loved them all in exactly that order. The farm had been in the family since 1820—a little less than fifty years after Daniel Boone led the first settlers into the state—and had supplied a life for the Joneses ever since, as it would for the children. But the dogs provided Jace with his main pleasure—hunting. He prided himself in his dogs because he trained them well, and he had a reputation as one of the best hunters in the county. He did not train them with kindness.

"Ya gotta clout 'em now and then, boy," he used to say to his oldest son when trying to hand down his skill in training. And there had been the times, when one of his dogs became particularly obstreperous, that the clout was delivered with the butt of a gun.

The training processes required patience, and Jace had a whole well full of that virtue for any dog that showed promise.

That fall, after harvest was done, Jace had taken five of his dogs out to work them and a friend from a neighboring farm had come along with three of his own dogs. The crisp morning held hopes for good hunting.

As Jace let his dogs out of the kennel boxes on the back of his pickup truck, he felt good about everything except for one of his dogs, an American pointer that had been the runt of a litter a year and a half before. The dog had never hunted, and seemed not quite alert enough for Jace's standards.

The pointer ran to some tall grass where the other dogs were relieving themselves, and squatted while he urinated. Jace yelled, "Now, look at that damned dog. He don't even know how to pee right. You wonder why I call him Clum? It's short for clumsy."

For the next couple of hours the two men moved through the woods, encouraging the older dogs to lead the younger ones through their tracking, and in holding point. Part of field work involves training the dogs not to shy when guns are fired, but Clum repeatedly squatted to pee when a gun went off. That habit did not sit well with Jace.

Just before lunchtime, when the two men were beginning to get hungry, Jace called all of his dogs in and leashed three of them to the bumper of his truck.

"What are you doing, Jace?" the other man asked.

"'Fore we go in, I wanna give this dumb one another run with old Brownie."

"Forget it, Jace," the friend chided. "Any dog that pisses—"

"Okay!" Jace cut him off. "But I jest gotta give him one chance. Hell, I been feeding this mutt for a year and a half. That's a lot of money."

The friend nodded; he knew that even purchasing huge quantities of bulk dog food can run the value of a dog up. An active fifty-pound hunting dog would eat about three pounds of feed a day, and that could run well over a hundred fifty dollars a year, even using the cheapest possible product available. Jace's pack of eight dogs was costing him a considerable price just to feed, and he had to add various medical fees on top of that figure.

Jace's favorite hunter was a seven-year-old black-and-tan coon dog who had been used for several seasons as a guide for the other animals. Jace swore he could tell old Brownie to go after rabbit or squirrel or raccoon and the dog would only seek that particular game; he had proven the boast on several occasions. Now he told the coon dog to go after a rabbit, and he ordered

Clum to follow. Several thrown rocks were needed to get Clum moving in the general direction, but as soon as he saw the coon dog taking off across the field, he joined in on the chase. The dogs disappeared into a woods on the other side of the field.

Jace gave them a couple of minutes before whistling a command to return. In seconds the coon dog appeared, but Clum was nowhere in sight.

Jace and his friend waited a full twenty minutes, but still Clum did not return. Normally the next move would have been for Jace to trek off, probably with his trusty coon dog, to search for the missing mutt.

Jace turned to his friend. "Let's load 'em up," he said.

"Ain't you even gonna look for him?"

"Naw, shit, he ain't worth the food I been givin' 'im. Let's get the hell outa here 'fore he comes back."

So Clum, a reasonably pleasant dog who seemed to lack the instincts for hunting, was left on his own, to wander around northwest Kentucky.

When it came time for him to fend for himself, he made out well enough. Like most dogs who are forced to do their own food gathering, he lost a couple of pounds, but he also managed to build up his body fiber on a diet of small wild animals and the plant life that was available for roughage. He roamed north, crossing the Ohio River at Grahamville, and he wintered in the Shawnee National Forest in Illinois. He made it over the Wabash River at New Haven without paying a toll and he even scrounged half a sandwich from the toll taker. At a truck stop on Interstate 64, a driver who loved hunting lured him into the warm cab with a hot dog, and he slept halfway across Indiana. Then, near Louisville, the driver took him out to walk in a field, and Clum kept walking.

By mid-spring he had traveled into Brown County State Park and, on one rainy night, he came across a cavelike shelter where he was to find some new companions.

Chapter Eleven

"Mary Ellen, do go get dressed."

Mary Ellen wiggled down into the cushions of the huge sofa as if she were trying to make herself less visible.

Her parents were in the library, having their nightly martinis and watching the television news before dinner. She saw no reason to dress for dinner because it was going to be just the three of them at home. Her mother had rejected the idea of going out for dinner because the rain would probably ruin her hairdo, and that would have negated her thirty-dollar investment and the tiring trip she had made up to Indianapolis that afternoon.

"*Mary Ellen.*" Her mother's voice had the lilt of a coffee grinder. "Now, dear, you *must* dress."

Mary Ellen plunged her eyes back into the reading assignment her psych professor had insisted must be completed by the time the Easter recess was over: "Hence, down to Hades! where our Queen has hastened on, . . ." *God,* Mary Ellen thought to herself, *could Goethe have imagined her mother when he was writing* Faust?

"Mary Ellen! For God's sake, get dressed." Her

father's voice was booming from the library now, and this command was one to be obeyed.

She swung her feet down from the back of the sofa and stared into the blazing fire. Anticipating another burst of directives from the library, she called, "Okay, I'll only be a few minutes."

She took advantage of the delay thus gained to enjoy the solitude of the fireplace for another couple of minutes. She enjoyed times like this: rain splashing against the sliding glass doors leading to the swimming pool; the peace of the evening; the pleasure of not having to do anything.

"Get dressed!" her mother ordered from the doorway. "Dinner will be ready in a few minutes and I'm not having you come to the table dressed in *those* rags."

Mary Ellen faked a friendly smile and replied, "Right, Mummy," as she jumped up and moved quickly across the living room.

Mary Ellen had given some serious thought to moving out of the house and into an apartment to escape just this sort of annoyance. She never had thought her mother really cared for her, and the many affectations in the life of the Crawford family were annoying. Dinners. Social dances. Church teas. All a bursting bore.

But where could she get the money to live alone and, especially, where could she find an apartment that would come anywhere close to the real comfort she had here? Her suite in the Crawford home was bigger than most apartments occupied by her college chums in pairs and groups.

She walked down the hall to her rooms. Inside her dainty parlor, she leaned tiredly against the wall, the aches in her legs and shoulders pleasant reminders of her first day back in the saddle.

The day had not been all that bad, she thought, *except for the incident with the bunnies.* The image of the dead animals brought Mel Ryan into focus in her mind, and she wondered again if she had been completely fair. Was he callous and cruel about dogs? Or, was he just upset? She decided she would go riding in the park again the next day.

She looked through the door leading to her bedroom and spotted Duke curled comfortably on her bed in his customary position, with shoulders pressed against her pillows and head aimed at the door.

"Duke!" she yelled as she ran at the bed. She dove the last three feet and landed in the middle of the large bed as she flung her hands out to grab at his paws. At the last possible instant, Duke flipped clear of her grasp, jumped over her body, and landed in a sitting position near the foot of the bed. She grabbed at his legs with a loud "Gottcha!" and Duke twisted free, leaping back across her and falling happily into the mound of pillows. Mary Ellen jerked herself up onto her knees and lunged at Duke as she emitted a gruesome sound that she supposed passed for a growl. Duke playfully tried to fend her off by waving his strong legs in the manner so typical of a boxer. She pounced on him with her full weight and he took the impact easily, sinking down into the pillows. He opened his mouth and clamped down on her right wrist. "Oohh-oohh," she whimpered as if mimicking a damsel in distress, and Duke instantly opened his jaws while purposefully forcing her arm clear with his wet tongue. He twisted his head around to study her eyes and saw she was not injured, and that the routine of play was going according to the tradition structured over the years. He wiggled out from under her weight and escaped again to the foot of the bed. The frolic lasted for another couple of minutes, then she slid to the floor and said, "That's enough. I've got to get dressed."

The bedspread was a rumpled mess, but Duke found a spot to settle down, to wait and see if the game could be resumed later. Mary Ellen selected a lightweight dress and fresh underwear, and went to take her shower.

It was a simple, routine dinner for the Crawfords: tomato aspic, smoked salmon, coq au vin, and chocolate mousse. Mrs. Crawford's climb up the social ladder of Columbus's gentry would never be hindered by a poorly set table. Mary Ellen was bored, and was just about to launch a major argument against her mother out of sheer perversity when she was rescued by the maid, who announced that she was wanted on the phone. She bolted from the dining room.

It was Mel Ryan.

"Are you busy tonight?" he asked after a quick hello.

"How did you get my number?"

"There are only nine Crawfords in the directory; your number comes right after Howard and right before Opal. I'm afraid there are five Crawfords who had their dinners interrupted, but I just wanted to see you again. Are you free tonight?"

It was one of the nicer things that had happened to Mary Ellen recently. Her social life at Indiana University had tapered off since she had broken with the group that released the dogs, and she was just as glad it had. But here was a guy actually seeking her out, and it gave her a warm feeling. "Sure," she said, "I'd love to get out."

"I can be there in fifteen or twenty minutes," he offered.

"Great," she said, "It's 520 Tipton Lane. Just before you get to the North Christian Church. Okay?"

"I'll find it," Mel said, and eagerness showed in his voice.

Mary Ellen sat there for a few minutes, bemused but pleased. She decided that she had been wrong about Mel's attitude toward dogs; she must have misunderstood him.

She was halfway back to the dining room when she heard her mother's raised voice: "Just stop it, Ken. I hate that."

Mary Ellen did not have to see what was going on; she already knew. Someway Duke had slipped into the dining room, and Ken was giving the dog scraps from the table, as he always did. She walked on into the dining room, and sure enough, there was Duke, drooling for all he was worth down at her father's end of the table.

"Stop it, Ken." Mary Ellen's mother sounded as if she might become hysterical.

Her father grinned and tossed a bit of roll to Duke.

Mary Ellen saved the moment. "I'll take him out," she said, and then ordered, "Come, Duke."

She snitched a scrap of cold salmon from the table and used it as a temptation for Duke, who followed her from the room.

Outside the kitchen door, in the graveled area between the house and garage, Duke suddenly stopped in his tracks. The evening was being sprinkled by a spring rain, and the air was clear and fragrant.

"Come, Duke," she ordered.

But Duke stood there, feet planted firmly and muzzle raised inquisitively, sniffing.

"Smelling one of your lady friends, Duke?" Mary Ellen asked in an amused tone.

Duke made a move to trot off in the direction his nose was pointed.

"Oh, no," she challenged. "Not tonight."

But Duke did not stop. He bolted into a run before she had finished speaking.

"Duke! Heel!" she shouted, giving the one command he usually obeyed. He skidded to a halt in the driveway.

Before he could reconsider, Mary Ellen ran across the gravel and grabbed his collar.

"Come on, boy," she said softly. "No escapades for you tonight." As she began leading him to the kennel around the side of the garage, she was totally unprepared for what happened next: Duke began balking.

The boxer has been bred so that his hind feet dig in to give power. While he gets no impelling power from his front legs, they still are heavily muscled. When Duke planted all four paws, it was all Mary Ellen could do to move him. She was forced to virtually drag the dog toward the open kennel door. When she was about five feet from the opening, an unfamiliar growl began seeping out between his clenched jaws, a thing that had never, ever happened.

She scolded harder and gave a sharper tug on his collar. He began to buck and thrash, trying to get free.

The growling increased. She was barely able to hold him, and thought of calling her father. But then, keeping her grip on Duke's collar, she grabbed a handful of loose skin at the back of his neck. Setting her feet, she threw her whole hundred and ten pounds into one big heave that carried him past the kennel gate. She slammed it shut, but before she could slip the bolt, Duke threw his whole weight hard against the wire mesh and it was all that she could do to prevent him from getting free.

As soon as he heard the bolt slam locked, his whole demeanor changed. He *grruuufh*-ed in his playful way, he bounced puppylike around the penned-in area, and he stood against the fence while emitting a loving whine that begged to be released.

Mary Ellen was furious. Not only had he kept her out in the rain much longer than it should have taken to kennel him but his behavior was something she had never experienced before.

Duke pressed his massive head against the restraining links of wire and presented his tongue in his best remorseful manner.

She remembered how he had growled at her and she said, "You shit. You can stay there."

Her dress was dripping, and she would have to change quickly before Mel Ryan arrived. But she was stopped from leaving by a mournful whimper from Duke. She went back to the gate, opened it, and went inside. The dog was all over her, tail wagging and tongue slurping affectionately. She stayed there petting him and making comforting sounds until he calmed down.

"Why did you act like that, Duke?"

When he looked at her, with head cocked to one side, she gave him a hug. "I know," she said. "You know you were a bad boy. That's okay. Go to bed now."

Duke seemed to sense that he would not be allowed out of the kennel for that night, and he followed her order, climbing into his warm, dry dog house. Mary Ellen left him and went to change clothes.

Chapter Twelve

AHA AMERICAN HUMANE ASSOCIATION

ANIMAL CONTROL STUDY

Computer Control Center, Denver, Colorado

EXHIBIT NUMBER 18-0-2C

DOG PARTICULARS
BREED Boxer
COLOR Brindle, black mask
WEIGHT 63 lbs HEIGHT 26"
NAME (IF KNOWN) Duke
AGE (APPROX.) 7 years

ANIMAL CONTROL REGISTRATION

Columbus, Indiana

CITY (COUNTY) STATE NUMBER

TYPE OF LOSS Uncontrolled pet.

The Pack had formed; only Duke was missing as the dogs began milling around the lair.

They had established a loosely defined territory that functioned adequately for their needs. Shelter was provided by a wedge of rock that had eroded and fallen across a small gully hundreds of years before. The formation had created a cave and at the same time had diverted the creek, which now flowed only twenty feet away and acted as a water supply. It cannot be said of them that they chose particular directions to travel or that they were motivated by some primeval ordinance to band together at this specific location. They had moved independently, and had accidentally happened on this grouping. Along the way to this juncture, they had each encountered adventures common to dogs in an uncontrolled or feral state.

Beau and Mitzie had been the first to discover the shelter when Mitzie had playfully chased a terrified chipmunk into the large opening. In the dark recess of the cave, the chipmunk had escaped, but Mitzie had stayed. The shelter seemed to offer security.

The den was dark but it was not damp because the rock-ledge roof managed to act as a watershed. Strong winter winds had blown in leaves and pine needles, and the floor was carpeted and soft. But the comfort and security of the cave were not the only inducements to settle; Mitzie was beginning her first season.

A bitch usually enters her first heat period at between six and eight months, but Mitzie was a few weeks early, which was not unusual for a dog living in the wild. Her proestrum cycle—the nine day period preceding fertility—produced a noticeable swelling of her external genitalia and a slight bloody discharge. Mitzie noticed the change and began trying to clean herself. She also felt rather lethargic, so she was quite pleased to settle in their newfound cave.

Beau also began to sense a change, and he experienced a confusing stirring in his own loins as the proestrum scent began to emanate from his companion. It was that attractive odor, too, which brought their first visitor.

It was very dark the night Clum arrived: Rain was falling, so Beau and Mitzie had curled up in the back of the den and were sleeping. To a human, the presence of Clum at the opening of the cave would have seemed silent and unnoticeable, because the pointer had eased up stealthily and now stood there, immobile, as he tried to calculate the source of that odor. But, to Beau, the arrival crashed into all of his senses and he sprung to his feet.

Beau was what would normally be called a slouchy dog in judging circles, and it is doubtful if he could have ever been encouraged to stand proud or respond gracefully to a handler. But, when meeting another male dog, Beau stiffened his back and snapped his head erect, with ears set high and nostrils flared. Even though it was raining, there was enough light available for Beau to see that Clum had approached the territorial limits. Beau moved forward slowly, taking small steps so that he constantly kept his leg muscles ready to leap. There was no menacing growl as Beau drew close to Clum, because the pointer was staying just outside the territory, and Beau was not attacking.

Beau did not go directly to Clum, but stopped first at the edge of the cave opening and urinated on the wall. He then passed closely in front of Clum and repeated his property claim on the other wall. Clum moved to the first spot, executed his urination, and then Beau quickly returned to resaturate the spot with his proprietary odor in that universal ritual of the dog world. Just how this ritual works is not fully understood, but it is known that the urine carries a specific enzyme that is uniquely significant to the dog's olfactory system. When Clum allowed Beau to have the last squirt, he also acknowledged that he was not intent on staking a claim of his own.

That established, Beau permitted Clum to have access to the cave, but when the pointer began to move toward the back of the den where Mitzie sat waiting, Beau withdrew his permission with a decisive growl. Clum retreated to the opening of the cave and bedded down for the night. The two dogs were about equally matched as far as size was concerned, so it was probably just the natural protocol of the canine world that moved Clum to accept both Beau's claim to the cave and his right to Mitzie. Beau slept fitfully that night.

The next day, his status in life changed entirely.

Jake, the Doberman guard dog, and Blackie, the harrier rescued from the vivisection lab, had been having an exciting time during the past few months. Within the concept of the feral dog, where the animal moves away from dependence on man, Blackie was developing true independence most rapidly. The harrier is one of the lines of dogs longest identified by man, and references to the breed may be found in the writings of the ancient Greeks. Packs were established for hunting in England as far back as 1260, and they were some of the first sport animals imported into the Colonies. This ability as a hunter is what has made the breed desired by man, but it also provides the harrier with a capability to survive in the woods.

The Doberman, on the other hand, showed up as a breed only in 1890, when the German Breeder, Ludwig Dobermann, combined the valiant strength of the Rottweiler with the stamina of the black and tan terrier, then mixed this combination with the smooth-haired

German pinscher to come up with the dynamic animal known today. Such selective breeding has moved the Dobie genetically away from independence, but it has also created a powerful animal, well-equipped to survive on his own.

The day after the arrival of Clum, Jake and Blackie showed up at the den.

It was early in the morning, and Beau had gone out on his own to scrounge up some scraps of food. That he was not at the cave entrance when Jake walked up probably saved Beau's life, because by the time he returned there had been a tacit acceptance by Mitzie and Clum of the Dobie and harrier. The scrap that ensued when Beau tried to reassert his territorial imperative could not really be called a fight, because it was over too quickly. The newcomers executed an excellent high/low attack on Beau, as Blackie nailed the right front leg and Jake hit the back of the neck. Beau joined Clum near the mouth of the cave, nursing only minor injuries, and Jake became boss dog.

Because Mitzie's pre-conception season had another six of its nine days to run, the dogs settled for what can best be described as an armed truce. They united into a loosely organized cadre and took their lead from Jake.

Clum and Blackie, by that time almost totally feral, were finding sustenance within easy roaming range of the shelter. Jake oscillated between consuming small game killed in the woods and scrounging garbage on forays with Beau up to the lodge or at campsites.

Activities fell into a type of routine for the next several days. The dogs were now aware, through complex instincts with which Nature has provided them, that Mitzie was not yet ready for breeding. Beau was provisionally permitted to bed down with his bitch, but it was obvious that the arrangement was to be purely temporary. Hunting and garbage raids were carried on nocturnally, and long periods of sleep were taken during the day.

Brilliant colors were splashing off the blooming dogwood trees, and flowers blossomed along the bank of the stream. Ambient noises such as the frantic buzzing of newly hatched flies and the industrious tapping of woodpeckers were muffled pleasantly by the steady, soothing gurgle of the nearby brook.

One morning Blackie, always alert, sensed a presence in the opening of the cave. His eyes snapped open but he did not move; a low, rumbling noise came up from his chest through his clenched teeth. The warning sound reached the ears of the leader, and Jake's muscles were yanking him erect even before his eyes were half-opened.

Jake faced Duke.

The distance between the two dogs was well over fifteen feet, but even at that distance Jake's sharp eyes could tell that this boxer was big, broad-chested, and equipped with a fierce-looking mouth.

In Jake's life as a guard dog he had had little contact with other dogs, but he had seen a dog like this before, and the memory brought back unpleasant, nearly horrible, images to his mind. There had been a boxer at the training school where Jake had been educated in the techniques of dog versus man. He still remembered the painful jerks of the spiked choke collar, used to encourage the dogs to learn quickly. He remembered agonizing spasms when an electric cattle prod had been jammed against his penis sheath or around his rectum. And he remembered the boxer, the private possession of the brutal, shaven-headed owner/trainer of the kennel. When a dog lagged behind in the training, the owner would encourage viciousness by pitting the boxer against the malperforming student, often a puppy. The boxer always won, though the conflict was always stopped before the trainee was damaged.

But Jake was older and stronger now, and he was ready for the boxer. Without pausing to void himself of encumbering urine or even sucking in a sustaining gulp of air, Jake bounded forward. His leap carried him past Beau, who had risen to meet the intruder, and then over Blackie, who was jockeying his legs under his prone body in anticipation of an attack. Jake touched the floor of the cave and was instantly airborne again, hurtling at the boxer. Jake had calculated the distance of his leap so that all four of his feet would be securely on the ground as his head arrived in the region of the boxer's throat, and as he felt the last of his four pads make ground contact, he hooked his wide-open mouth around and down in a move aimed at Duke's vulnerable neck.

Jake's jaws slammed shut on air. Mysteriously, Duke had catapulted backward just at the final instant and left Jake standing there, snapping at nothing. His sharp eyes reoriented and picked up Duke's mocking form six feet away. A second attack, carried out in exactly the same manner, was rewarded with exactly the same result. The agile, powerful, sixty-three-pound boxer was toying expertly with the infuriated Doberman. Bred by the English for pit fighting, the boxer has an ability to whip his body clear of an attack with ease. Not once during his encounter with Jake did Duke poke his paws in the air, which looks playful but is the first warning of a boxer attack. This day Duke was on a frolic.

Jake paused for a moment, announcing his intentions with the meanest snarl he could develop. Duke responded with a slight *grruuufh.* Jake moved stealthily to the right, hoping to back Duke against a rock outcropping, thus cutting off an avenue of escape. Duke reacted by springing effortlessly up onto a fallen tree. Then he simply vanished.

Jake's pride and skill, and probably his leadership, were now on the line. He vaulted up over the log, came to a halt, and searched with his eyes. There, peeking from behind a wide oak tree, was Duke, offering a second playful *grruuufh* right at the frustrated Doberman.

Jake was joined by his Pack, who were now as confused as their leader. From some inherited sense, they fanned out in a half-circle in a manner used by their ancestors to trap prey. Duke scooted through the perimeter between Blackie and Clum without as much as a tooth or claw touching his smooth, tight, brindle-colored coat. He took a new position, on high ground past the cave and near the stream, and voiced another *grruuufh* to the stunned pack.

The spot Duke had run to was flat and offered a large area for maneuvering. Jake kept his hackles sharply raised and puffed out warning spurts of air from his flews, but now he approached Duke warily. The boxer did not display his hackles, nor did he snarl warningly. He pricked his ears and rolled his head casually, as if to announce that he was not being aggressive. In the ritual established since prehistoric times, noses were presented and sniffs inhaled without touching. That Jake had been outmaneuvered did not mean that he had been subjugated; he still held the posture of a leader and, as such, assumed the dominant role in the introductory process.

Duke was by no means intimidated by this Doberman, who was a full three inches taller and thirty pounds heavier. The boxer was simply looking for playmates. He had been attracted to the spot by Mitzie's fragrance, but on finding this pack of males, Duke was in a mood to frolic. Patiently he allowed Jake to execute the remainder of the indignities that are critical to the ceremony.

Jake took his time sniffing Duke's various private parts, making a slow, cautious circumnavigation of the boxer's entire body. Working his way back up Duke's side, he came to the ear, and it was here that the first physical contact came. Jake's alert nose explored against the skin. Duke skittered a couple of half-steps sideways and allowed his hackles to rise slightly, but he stood still and kept his jaw shut as a sign of friendship.

Jake approached again and picked up at the ear with his inspection. With this second incident of direct contact, things seemed to calm down. Jake moved back to Duke's muzzle, where he forced his nose against the dark mask and ample folds of skin. Then he took a slight step backward and submitted to an identical, reciprocal survey by Duke. At the end, when Duke brushed his cold nose against Jake's muzzle, there was an instant of final decision; then, quite without warning, Duke spun around in a full circle and came to rest with his forelegs extended and his head close to the ground.

It was extremely difficult for Jake. He hadn't done anything like that since he had been a puppy, but he did manage a variation on Duke's performance. Turning a quarter to the right, then back a half-turn to the left, he ended up exposing his left side to Jake; a demonstration of trust. Then he expelled a muffled *grruuufh,* dropped down on his forelegs, and allowed his docked tail the almost unremembered joy of flicking back and forth. Instantly, Duke popped up, ambled over, and bumped shoulders with Jake. The next several minutes were devoted to inspections by the other dogs; then Duke joined the pack for a rest period deep in the cave. Just before dark, they rose in unison and went to the creek to drink.

Then, as the sun settled down over the rolling hills of south-central Indiana, Duke felt it was time to start for home. But first he led the pack to a spot he had visited before, and to some really entertaining activity with a mother rabbit and her bunnies.

Chapter Thirteen

Mel Ryan turned off Washington onto Tipton Lane. He was beginning to feel uncomfortable. The homes were large, the yards neatly groomed, and the whole place glowed with the aura of money. In fact, for a town of its size, Columbus showed that it was unusually well off. The hundred-year-old courthouse downtown had been expensively restored to Isaac Hodgson's 1874 concept of what a baroque county seat should look like, yet somehow it did not conflict with Myron Goldsmith's idea of what a modern, glass-enclosed newspaper building should be, as embodied in *The Republic* printing plant and headquarters on Second Street. Mel knew it took money to marry such extremes into an acceptable impression, and Columbus had obviously spent a lot of money on its buildings.

As he neared the end of Tipton Lane, Eero Saarinen's spearlike spire rose over the North Christian Church, radiating an aura of affluence over that part of town. The whole feeling contrasted dramatically with Nashville just a few miles west, but Mel was glad he did not live in Columbus; Nashville was his kind of town.

The uneasiness did not dissipate as he turned into the Crawfords' driveway. He saw that Mary Ellen's expensive horse trailer blended in quite easily with the house and ground. As he stepped out of his beat-up old car, he unconsciously began dusting off his clothes as if to make himself worthy to enter that rambling house.

As he slammed his door shut, a clamorous barking came from behind the garage, but Mel could tell from the pitch of the bark that the animal was kenneled. There is a pleading quality to a dog yapping to be released so he can defend his master's home, and Mel's trained ear picked up the tone.

The front-door light came on, and Mary Ellen came out quickly. She had obviously been waiting near the front door for Mel to arrive.

"Don't get wet," he called.

"Hi, Mel. It's only a drizzle."

She approached him, smiling, but her greeting was lost in the loudness of Duke's barking.

"Duke! Shut up!" she shouted into the darkness. But the noise continued.

"He's only doing his job," Mel said.

Mary Ellen laughed. "He's spoiled rotten. He wants to get out. I've just got to scold him."

She led Mel around the garage to the kennel, and Duke increased the furor as the stranger came into view.

"Bad boy!" she yelled. Then, lowering her voice, she added, "This is Mel. He's a friend. Be nice." Duke kept on barking angrily.

Mel asked, "Do you keep him kenneled all the time?"

"No. That's what he's so angry about. He wants to get out."

"Let him," Mel offered. "He'll be okay once he sees I'm no threat."

"Not on your life. He'd scoot right into the house, and Mother is not in a good temper."

Duke's fury seemed to be increasing. He made a strong leap, slamming against the chain-link fence.

"We'd better leave him," Mel said. "Or else calm him down. He could hurt himself."

"I know he won't shut up," she said, her voice showing irritation. "And if he keeps this up, Mother will be really upset."

Mel asked, "May I try?"

She shook her head. "I think you'd better not. He's been in a foul temper tonight. He even growled at me. I think there's a young lady in heat somewhere in the neighborhood. He thinks he has to go courting."

"It will be okay," Mel said. He lifted the safety latch on the entrance gate and began saying the dog's name softly, mixing in other gentle phrases as he eased the gate open. "Easy, boy. Take it easy," Mel cooed as he eased his hand inside the opening. He made a loose fist and offered the back of his hand slowly to the dog.

Duke sniffed.

In just a few seconds, Mel had moved inside the kennel and closed the gate behind, and Duke was giving him the complete inspection, smelling his shoes and trousers. Mel stooped down as he kept speaking softly; in less than a minute, Duke executed a moist lick to Mel's hand.

Mel broke out in a broad smile and gave the dog a gentle cuff on the side of the head; Duke jumped with joy and quickly spun around, then gave Mel another lick, this time on his face.

"Pretty good, Mel."

Mel stood and said, "He's a nice dog. Let's give him the acid test. Come on in."

Mary Ellen hesitated a moment, but Mel's handling

of Duke seemed to indicate he knew how to deal with dogs. She opened the gate and walked in.

Duke raced over to her with a gleeful *grruuufh* as he wagged his docked tail. As soon as she had patted him on the head, Duke ran back to Mel as if to show her his new acquaintance.

"He seems to like you, Mel."

"Let's see how much," Mel said and he walked to where she was standing. Duke had been bounding back and forth in excitement, but as Mel drew close to her, Duke eased off his activity and watched. Standing beside Mary Ellen, Mel called quietly, "Come, Duke."

The dog hesitated, and Mel repeated the command with a bit more insistence. Finally Duke obeyed. Just as Mel stroked the broad, wrinkled head, he also reached up and put his arm around Mary Ellen's shoulder. Duke froze and looked up at Mary Ellen with his head cocked inquisitively. When no sign of protest came from his mistress, he pulled his head around and licked Mel's hand.

"Now," Mel said proudly, "how about that."

"I'm impressed, Mel."

Mel turned, and his face was only a couple of inches from hers. He asked, "Shall we really give him a test?"

Mary Ellen eased her face away, gave Mel a grin and said, "I think the testing is over. Let's go into the house."

When they left Duke in the kennel, he offered a sad, mournful whine. Mel suggested they take him with them. Mary Ellen shook her head. "I know that sound. He wants to get out and take off."

"You're probably right," Mel said as they walked in through the kitchen door.

Mel came from a fairly modest background, so it was

understandable that he was slightly awed by the Crawford home as they passed through the huge kitchen, into the dining room, and then on into the main part of the house. They found Mary Ellen's parents sitting in front of a lovely fire in the spacious living room, enjoying after-dinner drinks and coffee.

"Hi, young fella," Ken Crawford said as he rose and extended his hand.

Mary Ellen made the introductions, then told them about how Mel had gotten along with Duke.

"You know dogs, then," her father asked.

"A little, I've worked with them."

"What about that dog of ours, then. Some mutt, huh?"

Mel smiled politely.

"I'd hate to think how many pups old Duke has floating around these parts," Ken Crawford said proudly. He laughed.

Mel offered a noncommittal nod and avoided comment by accepting a cup of coffee from Mrs. Crawford. While Mary Ellen's mother poured, she chatted socially, attempting to subtly probe Mel's background. But Ken Crawford insisted on coming back to the subject of Duke.

"You have a dog?" he asked Mel.

"No, sir. I've just come here and I'm living in an apartment. Work keeps me away most of the day, and I couldn't keep a dog locked up in a room."

"Just the way I feel," Ken Crawford agreed. "I don't think a dog should be caged. He's got to have a free run. You're right, young man."

Mel wanted to avoid an argument, but he felt he couldn't let that pass. "I mean locking them inside alone,

Mr. Crawford," he said. "I don't think a dog should be allowed to run free."

"Sure, sure," Ken Crawford said amiably, "There are dangers. But you have to let them out . . . you know . . . so they can . . . well . . . do their own thing."

"You mean to chase bitches in heat? I'm sorry, sir, but I don't agree. A male dog should be controlled just the same as a bitch."

"Wrong," said her father as he poured another drink. "A dog like my Duke can do a lot of good by improving mongrel bloodlines. And think of all the kids who can witness the miracle of birth after Duke's gone out and done his bit." He smiled, but his tone was serious.

Mel shook his head in frustration. This was his first date with Mary Ellen, and he didn't want to get his initial meeting with her parents off to an awkward start. But he could not let it pass. "How about the miracle of death, sir?" he asked. "Do the parents take those kids down to the animal shelter to see the unwanted pups put to sleep? In this country, we spend about twenty-five million dollars a year to dispose of five or six million unwanted dogs, and God only knows how many more million are dumped off on country roads to be killed by cars or to starve to death. The lowest estimate of dog births is thirteen *hundred* an hour. That's every hour, every day, every year. It is a major problem in this country."

Ken Crawford nodded. "I know, but my Duke is pedigreed. That makes a difference."

"Sure it does. If you breed Duke to maintain the fine bloodlines he obviously has, then that's great. You'll probably find good homes for the pups. But if he just wanders out and impregnates any random bitch, then he is doing nothing more than contributing to a huge problem. If he's going to roam, he should be neutered."

"Hey, boy, watch that talk," Ken Crawford said with a warning hand gesture. "Nobody's going to go slicing up my Duke."

Mel wanted to argue the point, but he knew it was hopeless. He had seen too many cases of men who treated their male dogs as extensions of their own virility. It would be virtually impossible to convince such a man that a totally different set of rules—and problems—existed for instinct-driven animals.

Mary Ellen saw that Mel was looking for a way out, so she suddenly announced, "Mel is taking me to the movies, and we'll be late if we don't hurry off."

When they were in Mel's car and driving away from her house, Mary Ellen said, "Wow! You come on pretty strong with my father."

"Sorry," he said. "I just disagree with people who won't face the responsibility of owning a dog."

"How can you say that? Duke gets the best food, he goes to the vet for his shots, and he has a good home."

Mel pulled off to the side of the road and stopped his car. "Look," he said, "Why don't we drop this. It won't prove anything."

She sat there looking at him. He was different from the males she had known at the university. He seemed mature and she liked him. But she was perplexed. "Why do you have this *thing* against dogs?" she asked.

"I don't have anything against dogs, Mary Ellen. As a companion animal, a dog is great. We've taken a wild beast and turned him into a lovable, loyal creature. But we've abused our role, and there are just too many dogs, and too little control. And that can be damned bad." He sighed. "The problem with dogs begins with humans, and a good example is your father. It's nonsense to think that cute little pup we hold in our hands is not a potential danger. He is."

All of her life she had been conditioned by her father to see dogs as nearly human and deserving of special treatment. Now Mel was challenging everything she felt about dogs. If he was right, then what she had done at the university had been wrong; releasing those dogs had been bad. She could not face that. But she didn't want him to feel everyone was attacking him.

"Truce, okay?" she asked, her hands raised in surrender.

He smiled. "Okay, a truce. I do get a little tense about this. Let me tell you one thing and maybe you can understand." He related his experience in Vail the winter before, when he had seen the elk slaughtered by the dogs. He ended by saying, "So, you see, I have some reason to get uptight."

"They were pets. You're sure?"

"I'm sure," he said emphatically.

"And you *killed* them?" she asked."

"Yes," he said, his voice strained. "I had no choice. No choice at all."

She nodded her understanding, and they let the matter drop.

They did go to the movies that night at the Courthouse Center, and they had a drink at the Surrey afterward. They danced a little and even ended the night with a reasonably satisfactory kiss. Yet, the story of the elk kill did not leave her thoughts. She tried to imagine what it would really look like but, as she formed the pictures, her boxer, Duke, kept interjecting himself into the scene and her stomach would turn. She tried to put herself in Mel's place and wondered what effect such an incident would have on her life. It had been frightening, of course, but it had also lost him his job. Little wonder his attitudes toward dogs were confused, and that the whole subject made him somewhat tense.

As she prepared for bed, Mel Ryan remained a question in her mind.

Chapter Fourteen

The Pack waited.

It had been raining since nightfall, and trickles of water splattered down the mouth of the cave. But the shelter remained warm for the Pack because most of the members were moving.

Earlier, because the restaurant at the Abe Martin Lodge emptied its kitchen bins shortly after seven o'clock, Beau had made a food-collecting foray with Mitzie. The change in weather had dramatically cooled off the area, and the dogs even spent a few minutes rolling and romping in an area of matted pine needles. The pickings at the garbage bin had been adequate. Once the dogs had slipped through a loose board in the protective wooden fence, they ripped open the plastic garbage bags with practiced movements. Roast beef seemed to have been a fairly popular selection in the lodge's restaurant that day; the scraps, bolstered by the usual hamburger buns, overcooked vegetables, and overaged sandwich fixings, offered them a filling, if not a well-balanced meal.

They strolled back to the cave after eating. It wasn't as if they were sleepy; they had slept most of the afternoon. But there was a sense of anticipation in the

Pack, and they stayed close to the cave as if waiting. The experience the night before with rabbits had been exciting, and had left them keyed up. And all the dogs sensed Jake's obviously changed temper, the way he had accepted Duke without direct physical combat.

When Beau and Mitzie arrived at the cave, Jake was sitting in the entrance just clear of the droplets of rain. He greeted them by sniffing Mitzie and growling at Beau. The growl was not especially menacing, so Beau raised his hackles, dropped his ears back, and walked cautiously back to where he normally slept. Everything was normal except that Duke, the newfound member of the Pack, had not returned.

It was almost as if they were waiting for Duke's return before they left the cave.

The white-tailed doe was sheltered under the thick boughs of a conifer tree. She was enveloped in that lethargy Nature seems to permit all of her creatures—the exhaustion of delivery combined with the euphoria of parenthood. It had been a strange noise that had initially snapped her out of her labor-induced sleep, but as soon as she had snapped her head up and perked her ears in search of the noise, she realized that her fawn was not curled in against her teats, nursing as it had been when she drowsily closed her eyes a short while before.

The doe could not really be blamed for falling asleep, because this was her first fawn. The irises of her huge limpid eyes were expanded fully, and she could make out some forms. The rain had stopped shortly before she woke but sparkles of water still fell steadily from the pine tree's branches in the weak moonlight. She looked across the slight mound of moss-covered earth where she had reclined while delivering her fawn, found the oak tree where she had eaten before her labor began. And there, by the poplar tree where she had scratched her back during the afternoon, front legs askew, short, white-tipped tail angled sloppily to the side, was her fawn.

The doe studied the fawn for a few seconds, savoring the full feeling deep inside her that was the universal reward for the pains of birth. Then she winced as she forced her hind legs up and rose painfully from her kneeling position. There was a hesitant, awkward motion to her slow walk as she approached her fawn, but she could feel her muscles loosen and her body grow stronger as she forced the activity. She nudged her fawn with her foreshoulder and the tiny offspring shied slightly and ended up with its uncontrollable front legs clumsily crossed. The doe eased her body around so that her hind legs were nearly straddled over the fawn's head. Instinctively it clamped its hungry lips to the offered nipple and began sucking. The doe moved her hooves in short steps that would not interrupt the feeding but would carry them both back to the dry shelter under the tree. She desperately wanted to lie down while her fawn finished nursing.

It was Clum who left first. He walked out of the cave, strolled down to the stream, and drank several small laps of water. Then he tried a trick he had learned long ago as a puppy running with other scent-trailing dogs: He forced water up his nostrils, then gave a hearty snort, increasing his sense of smell. Staying down by the streambed, he raised his head and sniffed the air. He sniffed several times, then trotted back to the cave and leaped up to the top of the ledge covering the entrance. Jake followed, and reached the top of the ledge just as Clum confirmed his earlier findings down by the stream. It was there all right, an odor that probably none of the other dogs could recognize.

Clum gave a high-pitched bark, and Blackie was out of the cave like a shot. Beau and Mitzie followed to where the Pack was gathering.

With Clum and Jake leading the way, they set off in a southerly direction into the wet, dark night. They had traveled three miles before they reached the spot, well inside the northern sector of Hoosier National Forest.

The doe had just lowered her head to raise a supple branch so she and her fawn could pass under the tree when she heard the strange noise again. She had heard other sounds when a careless squirrel would disturb a broken branch, or the noise of another deer as it scratched the ground in search of nuts or berries. But this noise was different.

The doe snapped her head up and checked the air currents for odor, calling on her special ability to recognize one odor-bearing particle in ten million parts of air.

The doe's beautiful eyes widened into ugly black pools of fear as she spotted the two orange pinpoints of reflected moonlight. Her milk-giving glands convulsed, yanking the nipple out of her fawn's mouth. At the same time, she raised her right rear hoof and nudged the fawn in under the branches of the tree.

She had nothing to do with the cry that escaped from her mouth as Clum made his sudden lunge out from under the cover of the bushes by the oak tree.

The doe reared slightly, pulling her front legs off the ground and swiping out with her razor-sharp hooves at her attacker. But Clum was an old hand at such encounters. He dodged easily and intentionally banged into her rear flank, hoping to tilt her off balance. Jake, just behind Clum would have been sliced by the doe's hooves had she not been staggered by Clum's body check. Jake's first pass was a dive for her neck, a natural attack for him but an easy thrust for the doe to counter by gracefully twisting out of the way. The Doberman managed to inflict only a slight puncture with one of his canine teeth as he snapped his powerful jaws shut.

Hoping she could draw the enemy away from her fawn, the doe sprung forward with a powerful leap that carried her over six feet from the branches of the tree. She was just getting set for running when she felt a searing, burning sensation on the upper part of her right rear hip. Blackie had jumped high and sunk his teeth into her, and he had torn open a large wedge of tendon and muscle.

Clum had spun around by this time, and he made a long vault that carried him all the way to the doe's left hind leg. His momentum carried him past the leg, but he was able to clamp his teeth tightly so that they acted as an anchor. His own body twisted but he held on tenaciously, jerking his head in an effort to tear her deep leg muscle. In panic, the doe arched her body forward, threw her rear up and kicked her hind legs. Clum landed five feet away, momentarily stunned.

As the doe's rear legs returned to the ground, she felt the impact of Jake's body hit her back, then his teeth sink into her neck. In blind panic, she bolted forward, trying to escape. Accidentally she ran under a low branch, and Jake was dislodged with a painful howl and a gash just above his right eye. As she bolted, Blackie made a running pass at her stomach, and sliced open an eight-inch gash in her belly.

The cry from the doe was nearly human, but it was the diverse sounds of barking, growling, and snarling that tempted the fawn to come out from under the tree. The bandy-legged tiny deer emerged, and the first dog to spot it was Mitzie.

The beagle bitch and her most faithful companion, Beau, had only contributed noise to the excitement with the deer; they were both wary of the extremely large-looking doe. But the fawn was only slightly larger than Mitzie.

With Beau following closely, Mitzie bounded up to the fawn and grabbed it by the throat. Having no experience at defense or even play, the fawn twisted and accidentally hit Mitzie's ear with its hoof. Mitzie yelped in pain, and Beau charged, hitting the fawn behind the ear and tearing skin. In a fury, Mitzie made another lunge at the fawn's throat. This time she merely managed to get tangled in her own ineptness and fell to the ground right in the path of the doe. A hoof in her midsection brought a second yelp. Beau grabbed the fawn and shook his head furiously, but the newborn deer's bones were flexible and the damage was slight—until Beau slammed it to the ground and broke one of its legs.

While Beau turned his attention to the whimpering Mitzie, the fawn somehow managed to drag itself back under the tree. It struggled on past where it had been safe and out the other side of the haven. Ten feet farther on, it fell over a ten-foot ledge and broke another leg.

Despite its extreme youth and motherless state, the fawn clung to life a full day and a half before it died in the ravine.

Its mother was luckier.

Enraged by the pain of the gash over his right eye, Jake plowed back into the brawl and joined Blackie, who was trying his best to tear the entrails from the deer.

Clum had revived and was gnawing on the doe's rear legs; their prey had been paralyzed into inaction by the strike at her stomach.

Weakness from giving birth, loss of blood, and strength-draining pain finally brought the doe down. The dogs bit and tore at her until, mercifully, she passed out and then died.

Blackie and Clum were the only ones who ate any of the body meat before the Pack moved on.

Within two hours, their blood lust not satisfied, the Pack stumbled upon the chicken farm of Luke Amoury, about a mile outside Hoosier National Forest. That was where Blackie was shot.

Chapter Fifteen

Greasy Fowler yelled out into the front part of the office, "Hey, y'all take care of things. I don't want no calls and I don't want to be bothered."

Greasy knew that being a sheriff in Brown County meant getting re-elected every four years, and Luke Amoury was the patriarch of one of the biggest families in the county as well as the richest farmer in the area, which meant both votes and money for a campaign.

"Now, Luke," Greasy said. "What can I be doing for you?"

Luke Amoury was angry and it showed. "They killed some more of my chickens last night, Greasy."

"They?"

"You Gawddamned right. A whole damned pack of dogs."

The sheriff frowned. "A pack? Now, come on, Luke. We ain't had a pack of dogs around here since hell knows when."

"We ain't, huh? Well there must have been five or six. Maybe more. They killed ten birds before I could git my gun." He shook his head. "I only got one at them."

Greasy looked puzzled. "You want me to send someone out to see if we can identify the dog, Luke? You know that ain't too likely."

Amoury shook his head impatiently. "You don't have to come out. I got the damned dog in my truck. I want you to see something really weird." He stood up and started toward the door.

Reluctantly, Greasy got up and followed.

As they walked out of the building and went to the parking lot facing the alley, the sheriff wondered about Luke bringing in the dead dog. Dogs that farmers shot were normally used as carrion to keep other predators away from chicken coops and pigpens.

They approached Amoury's mud-splattered pickup truck, and Luke said, "There, now. Look at that."

Greasy Fowler leaned over the side of the truck body and looked down.

Luke's truck was about normal—a layer of chicken droppings, a busted ax handle, coils of wire, pieces of rope, and lots of dirt. The only thing that made it different was the dog.

Greasy pulled back in disgust and said, "Luke, you lousy bastard."

There, lying in the filth on the bed of the truck body, was Blackie, his right rear foot shot off at the hock, a rusty piece of bailing wire twisted as a painful tourniquet, and a bloody, oil-stained rag tied as a muzzle around his mouth. Blackie was alive but weak; he lifted his head in pain, trying to free his mouth.

"Why the hell didn't you finish him, Luke?" the sheriff said in disgust.

Luke was not fazed. "Look at him! I didn't shoot him 'cause I wanted you to see 'im. Look!" He pointed to unfamiliar bumps and lesions on Blackies' stomach and side. Some of the Teflon sutures implanted in Blackie at the vivisection center had become infected, others had partially fallen out and dangled down like white plastic barbs.

"Okay," he said impatiently to Luke, "Okay I've seen them." He spun on his heel and headed back toward the entrance to his department.

Luke came along behind, trying to explain: "If I'd'a killed him you wouldn't've seen them *things.* Why are you takin' on so 'bout a lousy cur dog, anyway? You think he worried about hurtin' my chickens?"

The sheriff went inside and stormed up to the dispatcher's desk, with Amoury right behind. A deputy sheriff was filling out a report form at a desk, and Greasy ordered, "You go out into the parking lot. There's a dog in the back of Luke's truck. Take that dog over to the vet and have it put to sleep."

The deputy grabbed for his hat and started around the counter.

The sheriff added, "The mutt's got some *things* sticking in him. Ask the vet to look at them and tell me what they are. Okay?" He turned to Luke Amoury. "Okay. We'll see. I'll call you when I hear something."

Luke stood there, glaring. He said, "You got something botherin' you, Greasy?"

The sheriff took a deep breath, remembering those votes. "No, Luke. I'm sorry, it just hit me wrong. I'll get on this right away and see if we can locate those dogs. I'll call you."

Luke jammed his hat back onto his head, said, "I'll be waiting," and left the sheriff's office.

An hour later Greasy stopped by the vet's office.

Dr. Clayborne Pierce was the only vet in Nashville, and still he did not make out very well. Most people still took their animals up to Bean Blossom, six miles north of town, because old habits die slowly and Dr. Pierce had opened his office only three years before. But Clay Pierce had done several favors for Greasy Fowler, and the sheriff sent quite a bit of work to the new vet.

"It looks like a suture," Dr. Pierce said. "Some kind of a plastic device."

Blackie's body was on the examining table. The vet had ended the dog's pain with a quick-acting shot of pentobarbital sodium, and the postmortem examination had been completed by the time Greasy had arrived. Blackie had more than his share of diseases. He had three easily detected parasites in the form of internal worms, he was harboring a distemper virus, and somewhere he had picked up a substantial dose of leptospirosis, a syphilitic germ passed in urine.

"If it wasn't for the sutures," Dr. Pierce told Greasy, "I'd say this mutt had never been cared for by man; he is really in bad shape."

Greasy nodded with a tinge of impatience. "But what the hell can I tell Luke Amoury, Clay? He's on my ass something awful."

"There isn't really much to tell. I'll run a scatology on him if you want. That's all I can offer, Greasy. It'll take me about an hour to do a complete analysis. I'll call you."

"A *what?*" Greasy asked.

Clay Pierce gave a friendly chuckle. "I mean, I'll open his intestines and run a check on what I can find. I already checked the stomach and found mostly chicken meat and feathers. But there is some other matter in there, and when I'm through I might be able to give you a pretty good rundown on what this dog has been feeding on, which might give you some clue as to where he's been. What more can I do?"

Greasy pleaded, "Make it fast, huh, Clay? Old Luke's raisin' hell. Claims there's a pack of dogs runnin' loose."

Within half an hour of Greasy's departure, Dr. Pierce had discovered enough to raise one more important clinical question. He needed another hour to confirm his unpleasant suspicions.

Chapter Sixteen

Earlier, when Mel arrived at work that day, the superintendent's secretary told him to meet Oscar at the Family Campgrounds.

Mel nodded, got back into his car, and drove the mile to the campground. He spotted Oscar's pickup truck pulled up alongside one of those expensive camper vans.

As Mel parked, Oscar left the man he was talking to and walked quickly over.

"What's up?" Mel asked.

"That man there," Oscar replied, indicating the man standing by the motor home. "His name is Jeb Lewis. Came into the park yesterday from Chicago for a week's stay. He's got a wife and two kids, and he *had* a dog, until six o'clock this morning."

"So?" Mel said. "I'm sure lots of dogs wander off for a couple of hours. I know it's against the rules, but why the fuss?"

"The fuss is," Oscar answered impatiently, "that I got the chief ranger coming over from Hoosier National Forest in about fifteen minutes. They had themselves a deer killed by dogs last night."

"Damn," Mel said softly. "Do you think it ties with the rabbits I found yesterday?"

"I sure hope not."

"Oscar, why don't I take the jeep out and have a look around," Mel said. "If nothing else, I might locate these people's dog. And it there is a pack—"

"No!" Oscar interrupted. "You talk to the man and do what you have to, but no going out looking for some nonexistent pack of dogs." Then Oscar softened as he added, "Leastways not until I talk to the ranger. Let me see what he has to say."

Mel asked, "You going to tell the ranger about our rabbit kill?"

Oscar repeated, "I hope not. Not if he don't ask."

Mel smiled. "Okay. What do you want me to do?"

"Well," Oscar explained, "if you got some magic way, I'd like you to go get that guy's dog back where it belongs." He indicated the man standing beside the motor home.

"You're really upset, aren't you?" Mel asked.

"Hell yes." Oscar paused. "It ain't this dog. It didn't even get out 'till this morning, so it wasn't after the damned deer. But I hate talking to them rangers . . . "

"Me, too," Mel said. "I've had my fill of them. I'll talk to the guy, and maybe his pooch will show up before your meeting."

Oscar nodded, waved to the man by the motor home, then left in his pickup truck. Mel approached the man by the camper. He'd hardly finished introducing himself when the man launched into a harangue about how good his dog was. Mel listened to the typical speech; he had heard dozens over the years he had worked with dogs and dealt with owners.

"Yes, sir," Mel tried to cut him off respectfully. "What's the dog's name?"

"Vickie," the man answered, "It's really Victoria but we call her Vickie."

"Could she be in heat?" Mel asked.

"Nope. We showed her for a while, then it became too expensive. We had her spayed about a year ago."

Mel was pleased to hear that. If the missing dog was not roaming because of a natural urge, she might show up at any time. Also, her scent wouldn't be attracting every dog for miles around.

Mel asked for four portions of the dog's normal food. Then, with an armful of Gaines-burgers, Mel thanked the man, said he'd be back in touch, and finally asked that the park office be advised when Vickie showed back home. He left in his car.

It took Mel about fifty minutes to drive to the different locations around the park where he wanted to place the food. Half of the time was devoted to giving instructions and listening to the same anecdotes about dogs that he had heard in one form or another for years. His last stop had been at the camp store, where he was sure the storekeeper's old brown mongrel would eat the food. It was still worth the chance.

With that out of the way, Mel went back to Oscar's office.

Oscar's first words were, "Don't ask. It was a shitty meeting."

Mel walked toward the desk, and his smile faded as Oscar slid a batch of Polaroid snapshots across the polished wooden surface. The photographs were gruesome. The front of the doe was chest down with her head between her front legs. The rear of the animal was twisted all the way around so that her tail was on the bottom and her stomach faced the camera. Her entrails had been clawed out and spread carelessly about the area.

"That's rotten," Oscar said, still feeling slight pangs of nausea. "The ranger claims they've lost sixteen deer so far this year. That's six more than the year before."

"Why do you seem so convinced there is no pack of dogs?" Mel asked. "Everything is pointing in that direction."

Oscar shook his head. "I run a good park, Mel. I've been doing it for a few years without major problems, and I think I know what is going on in my own park. We've never had a dog pack, and I don't see any real evidence that we've got one now. And I will *not* have people afraid to use this park because of some half-assed rumor about dogs. Hell, they're after me to raise the admission already. Think what this kind of thing would do to the operating funds."

"But, Oscar, it couldn't hurt for me to look aroun—"

"I said no, dammit! Now let's drop it and see if we can't get some work done around here."

As the door closed behind him, Mel stood and looked across the park, wondering where to begin. Brown County State Park is over five miles wide at one point, and seven miles long from its north gate to Little Blue Creek, which empties into Monroe Lake in the national forest. It was a lot of area to hide in.

At first he was inclined to go in the general direction of the forest, because the deer had been killed there. But he had heard that the Sassafras Audubon people had been checking trails in that area in hopes of getting the

Nebo Ridge declared a wilderness study area. Surely their presence would scare off any pack of dogs. The western part of the park was bordered with a reasonable amount of private homes, and there had been no complaints about dogs that he had heard of during the past few days. So, finally, he took the road that circled around Jimmie Strahl Lake. In a few minutes he was hiking through virgin woods toward the eastern boundaries of the park.

It was warm and damp in the woods; the bright sun was working on the moisture from the rain the night before, and the humidity made Mel's shirt cling to his skin. He pulled off his lightweight jacket and slung it over his shoulder.

He heard birds chirping overhead, but he kept his eyes on the ground looking for any sign of dog tracks. The undergrowth was heavy, with twigs and vines grabbing at his legs, but he moved easily from years of practice. The trees and branches broken from winter snows were already beginning to convert to humus, and the first spring wild flowers were battling their way up, reaching for sunlight. It was still cool in the dense cover of the conifers, which would not tolerate warmth over their shallow roots.

He moved as quietly as possible for about twenty minutes. He had found the almond-shaped indentations of the ground tracks of several deer, and there were beaver droppings as well as cracked corn shells that had probably been littered by a squirrel. But there was no sign of dogs. Either they were not there or they were maintaining an incredibly low profile.

He had been moving for nearly a half-hour before he

began to sense he was being watched. He froze, waiting to see if there were any sounds. The woods were silent, except for the normal sounds that should be there.

He waited several minutes, satisfying his need to move only by shifting his weight from foot to foot. He could see nothing, and there were no audible signs, but he knew there was something.

"Hey!" he shouted, but nothing reacted.

He bent down slowly and searched with his hand for a rock or a stick; he kept his eyes scanning the thick underbrush around him.

His hand closed over a rock that was small enough to throw and large enough to make a loud noise. He picked a random direction and tossed it hard into the bushes.

There was a scurry, and Mel headed in that direction. In less than a dozen steps, he came into a small clearing, less than ten feet across. It was empty.

He stood still again. He moved his eyes around the clearing, and then he found his first evidence.

The paw marks read like a child's book to Mel. There were the deep impressions where the dog had been standing, pointing in Mel's original direction. There were the shallower marks where the dog had apparently backed up a foot or so when he had yelled, and there were the squirm marks where the dog had jumped back when the rock had been thrown.

"Here, boy. Good dog. Come."

He was surprised that his calling had a ring of uncertainty, that his words lacked confidence. As he repeated his call, he heard the noise off to his left. But he was distracted from it because the bushes right across the clearing began to move.

A black face poked its way through some ferns, and Mel could see a lolling tongue and bright eyes. He felt his tension drain away.

He stooped down, because he had learned that dogs seem to trust man more if there is a less ominous size difference. Crouching, he said softly, "Come. Good dog." He gave a slight snap of his fingers, another of the universal man/dog communications.

There was another noise, this time off to the right. Mel heard it, but his prime attention was on the dog that was now moving slowly out of the clump of ferms. It was a mixed Labrador, and the dog showed he was quite willing to accept man: The tail was pointed nearly level and flicking rapidly back and forth in the kind of all-out way that moves the dog's whole hindquarters. The signal was clear; the dog was friendly.

Mel said, "Come on. That's a good boy." At the same moment more sounds to the right and left of him. He heard them this time, and he began to stand up. It was good he did, because Jake hurtled out of the thicket, his bulletlike head aimed right at the spot where Mel's own head had been. One of the characteristics of the breed is the Doberman's ability to hold in a warning bark until just before he makes contact during an attack. Jake never uttered a sound as his jaws noisily snapped shut merely inches away from Mel's head.

Mel's muscles were just beginning to tense for some kind of defense when a loud bark burst into his right ear, and Clum's jaws snapped on Mel's shirtsleeve. A glance at the pointer was interrupted by a flash of black, and from the corner of his eye he saw the Lab was running toward him.

The Doberman had landed several feet away after

missing Mel, but the dog was agile and he twisted in preparation for another leap.

Mel wanted to run. It was a natural impulse, but one that he knew was foolish. The Lab was not all that big, nor was he truly vicious; Mel thwarted Beau's lunge with a hard kick under the dog's chin. Beau fell whimpering and retreated just in time to get out of the way of Jake, who was launching his body at Mel's head again. The only weapon Mel had was his work jacket, and he pulled it from his shoulders and shoved it at Jake's face. The powerful jaws closed on both jacket and hand, ripping the cloth and inflicting a slight gash in the back of Mel's hand. Again Jake's momentum carried him past Mel, and he landed in a confusion of flapping cloth. Quickly Mel launched another kick, and this time he connected solidly with Jake's scrotum.

Normally, kicking at a Doberman could result in one sliced-up leg, but Jake's dangerous mouth was full of Mel's jacket, and so the gamble had been worthwhile. Yelping, Jake rocketed six or eight feet through the air and crashed down into the bushes.

Mel made his break.

There was one short time when Mel thought he was in trouble; that was when Clum came bounding after him and clamped down on his pants. Mel doubled his fist and belted the pointer in the side of his chest, and luckily the dog twisted loose.

Mel ran through the woods, back in the direction from which he had come. Small branches slapped at his face and briers tore at his clothing and scratched painfully into his skin. His lungs were aching, screaming for more air, but he ran with all of his strength.

The terrain was rough, scarred with gullies and

knolls that were heavily covered with trees and underbrush. He would vault a fallen tree only to find he was stumbling down a slope into a creek. And, as he splashed into the water, a wall of leaves and boughs would seem to block escape, until he forced himself to plunge through. He had run for nearly a full minute before he finally fell, his foot tangled in a vine. Unhurt, he decided to listen a moment to determine if he was being pursued.

It wasn't the heavy-throated bellow he heard that bothered him; that noise was coming from the pointer, who had been bred to excitedly announce a chase. Mel felt he could handle the pointer, but the noise that alarmed him came from off to the right. That sound, Mel knew, was coming from the powerful body of the Doberman plowing through the woods, probably on a course intended to intercept Mel's line of travel.

Mel thought of just staying there, not moving, hoping they would not locate him. He discounted that quickly. The Doberman would only double back, pick up Mel's scent spoor, and track him down. When the noise coming from Jake stopped, Mel knew he had to act, because the dog had stopped to listen for his prey.

Mel's college classes on animal morphology had stressed what little is known about territoriality. He knew that if he could just get away from the Doberman's territory there would be less probability of a continued chase. Of course, he couldn't know how far the Doberman's territory extended. But it was his only chance.

He jumped to his feet and began running down a gulley in a direction away from where he had last heard the Doberman. He ran for another minute, then halted.

There was silence. Even the bellowing of the pointer had ceased.

He did not intend to test his theory further. He moved as quickly as possible in the same direction for a full fifteen minutes before he paused to take a bearing that would lead him to his truck.

Mel drove fast as he circled Lake Strahl, eager to get back to inform Oscar of what he had found. But as he passed the cutoff that led past the Wild Mammal Exhibit, he spotted Mary Ellen's horse trailer and car down by the horse barn. He turned sharply and drove across the grass, then skidded to a halt.

As he walked toward her she waved, then saw the blood dripping from his hand and ran to him.

"What happened?"

"I met some of those friendly, playful neighborhood dogs," he said.

As she fussed over his wound, he told her of his encounter in the woods.

"So it is a pack, after all," she said. "She looked up at him suddenly. "Was Duke——?"

He was glad he could shake his head.

"My stupid father let Duke out last night while we were out," she said. "It was like he was getting back at you. I was so mad this morning that I could have spit. Then my dumb mother made me go to a church social . . . Oh, damn, I wish I could have been with you."

"I'm glad you weren't."

She had pulled off a scarf she was wearing around her neck and had partially cleaned his wound. Reluctantly he pulled his hand away. "I've got to get to the office, Mary Ellen. I don't think you should go out on the trails, though. That bunch of dogs could be on the move."

As he left her, she was loading her horse back into the trailer for the four-mile ride to the Saddle Barn.

He stopped by the park's camp store and asked the people there if they had seen or heard of the missing dog. They had seen no dogs at all, but he told them to warn hikers to stay out of the eastern section of the park.

In another couple of minutes he was at his office, but Oscar had already gone home for lunch.

"Call him for me, will you, please," he asked the secretary, who was munching on a sandwich.

Through a mouthful of bologna she asked, "Now?"

"When you can talk," he said. He went into Oscar's office and dialed the sheriff's office.

"He ain't here, Mr. Ryan. Matter of fact, he's on his way to the park."

"Now?" Mel asked.

"Just as soon as he picks up the vet; they're both comin'."

Mel cut off that call as the secretary said over the intercom, "Mr. Korteen on the line, *Mister* Ryan."

"Oscar," Mel burst out, "we've got dogs in the park. Now I know for sure."

Oscar's voice came over drowsily; it was his habit to take a short nap before eating his lunch. "Mel, what the hell are you talkin' about?" he said.

Mel fudged slightly, explaining that he had come across the dogs while he was checking an erosion problem. When he told Oscar where the attack had happened, Oscar exploded. "You walked all that distance to check on erosion? Bullshit! You and I got some talkin' to do, Mel. I'll be in right after I eat."

Oscar arrived in forty-five minutes, just five minutes ahead of the sheriff and vet.

As the sheriff walked in, he spotted Mel's torn and bandaged hand. "Boy, what hit you?" he asked.

Oscar interrupted. "Ain't no care of yours, Greasy. What are you doin' out here?" He nodded a curt hello to the vet.

Greasy told about Luke Amoury's chickens, and about the scatology the vet had performed on Blackie's body. The feces had been cooked on a glass lab dish and studied for content both under a microscope and by eye. Traces of fauna had been present, including field mice, deer, chicken, and rabbits; the flora consisted of leaves, grass, wood splinters of tulip poplar, and fragments of hickory nuts.

"So," Clay Pierce concluded, "The only place I can put all of those things together is right here. Right?"

Oscar's temper flared. "Crap! There are a dozen places you can find the same damned combination in this county."

The sheriff walked across the office and put his finger on the huge map of the area. "Look, Oscar, there was a deer killed in the national forest." He moved his finger from the forest to the Amoury farm location. "And Luke got a bunch of chickens ravaged by dogs, right? And the dogs in town were seen here and here." He indicated two places near Brown County State Park.

"So, they're nearby. So what?" Oscar was losing his temper. "There's lots of places around that have all that stuff Doc Pierce named. And they probably did pass through the park." He saw the look on Mel's face and stopped.

"Oh, shit," he said in exasperation, "Go on and tell them, Mel. I know you're damned glad to. Go on."

Mel hesitated, not pleased to be put in the middle. But he felt he had no choice. He told them the whole story.

When Mel finished, Greasy frowned at Oscar. "What the hell was you plannin', Oscar? Thinking of keeping it a secret?"

"I was plannin' no such thing, Greasy Fowler. But I know there ain't no dogs living in this park."

"Jesus," Greasy yelled, "What'd'ya need to prove it? You waitin' for them to kill somebody?"

"Now you just wait a goddamned minute," Oscar said, his face red. "You got no proof of any—"

Clay Pierce held up his hand. "Hey, let's not get into a pissing contest about things that are in the past. What are we going to do now?"

Everyone in the room looked at Mel.

It was a moment or two before he realized they were staring at him. "Hey," he protested, "don't look at me. Those damned dogs nearly whipped me. All I know is, they're out there in the park, and they're mean."

They talked for a quarter of an hour about what the solution might be. Finally the vet came up with a suggestion that satisfied everyone: "I'll bring over a couple of tranquilizer guns tomorrow and go out with Mel to where they attacked him. We'll try to find them again and bring them in to my clinic and see what gives. It won't take too long to see if they're truly feral. If they are, then you can make the decision, Greasy. How about that?"

The proposal seemed good and was accepted. They were just about ready to break up when the phone rang. Oscar picked it up, listened a moment, and handed the instrument to Sheriff Fowler.

The expression on Greasy's face was grim when he put the telephone down. He turned to Oscar.

"You're right, that the dogs ain't in your park," he said softly. "Not any more they ain't."

Chapter Seventeen

Duke had sensed the change just as soon as he crossed the narrow stream and approached the cave.

He had meandered on his trip to the park, sensing that he had been released too late to join in any activity for that night. He did not arrive at the cave until after midday.

The change was clearly visible to the eyes: Beau and Clum milled around the grassy area a few feet from the cave entrance where Jake lay prone, baring his teeth and emitting a nasty growl. There was another dog there, a well-groomed collie, but she was keeping her distance. Blackie was missing.

When Duke approached Jake, his customary *grruuufh* was greeted with a snarl. He backed off, joining the other dogs.

During the night, in addition to all the other excitement and activity, Mitzie had come into her conception phase of heat. Jake was not sharing. He was standing guard.

About dawn, after they had returned from killing the chickens and losing Blackie to the farmer's shotgun, Jake had responded to the new odor and receptiveness of Mitzie. She had accepted him, awkwardly and with some pain due to the difference in size. After the first encounter, she had slept for more than an hour; then, when she woke up and tried to leave the cave to join Beau, Jake had forced her back. She accepted him again, but again the mating was painful. She wanted to be out of the cave and with Beau.

She had slept during the period when Jake had led Beau and Clum out to defend their territory against Mel's intrusion. When the dogs returned to their base, they had rested. They woke just before Duke arrived.

Jake snarled again but remained where he was, with his legs set as if he were about to spring. The other dogs kept their distance.

Duke had been around enough breeding sessions to know Mitzie's new condition. But he had also come into the Pack with the acknowledgment that Jake was the leader. So, joining Beau and Clum, Duke paid some attention to the collie. A game of chase and nip developed among the outcasts from the cave, and Jake watched them with disinterested aloofness.

Mitzie was quite ready to join the activity but she was cowed by a fierce, guttural noise from Jake. She watched patiently while Beau attempted to mount the collie in a mock mating that had been provoked by Mitzie's odor. Then she pulled herself to her feet, when Jake's attention was distracted by the activity with the other dogs, she scooted speedily past him.

She had timed her exit with skill, waiting until Jake's rear had moved slightly off center and then passing on his right side, opposite the leg that had not been uncocked and less tense. Thus she was able to slip between Jake and the cave wall before he could leap up

Jake's roar and Mitzie's frightened yelp turned the

other dogs' heads in unison just as the beagle made it clear of the cave entrance. In a panic, she cut sharply to the right and ran for the stream.

Duke took off in hot pursuit, but in a few bounds Jake passed him with a menacing growl. The collie, Vickie, joined in the chase, probably thinking this was part of the group's routine playtime, and she was followed by Clum. Beau brought up the rear, attempting desperately to switch from a humping to a running mode.

Mitzie cut left at the stream and took advantage of the gradual rise of the terrain. She could hear Jake and Duke both splashing in the stream water, and she sensed that they could easily overtake her if she attempted to turn for high ground.

The chase carried them north toward the boundary of the park, which is where Mitzie swerved west to avoid crossing Highway 46.

She was able to stay ahead of Jake because he was slightly diminished in strength. Duke did not attempt to pass the enraged Doberman, and Beau, Clum, and Vickie were just able to hold up to the pace being set. Mitzie's beagle bloodlines gave her the stamina for a prolonged run, and she might have continued along the edge of the highway for miles, but as she passed the Salt Creek Shopping Center she was forced to dart across the road because two other dogs had been attracted by the barking and were running directly for her. With the whole collection of dogs fanned out behind her, she headed directly toward the playground of Nashville's Elementary School.

Chapter Eighteen

Bessie Carlton had ducked into the teacher's lounge off the playground to sneak a smoke.

She was enjoying her new assignment as acting principal and, although she did not wish the regular principal any complications with his appendix operation, she hoped the job would last for a while. She had managed to get the afternoon phys-ed classes started on time, so now she enjoyed her cigarette in relative peace.

She had smoked it halfway down when she heard the first scream.

The fifth-grade boys had been playing baseball at the east end of the playground, and the girls were at a confused game of volleyball on the slight rise toward the west end. The boys had heard the barking but it was only Mitzie they had seen as she scampered across what was loosely defined as left field. The gym instructor had tried to yell to the first baseman to let the dog go, but the boys' excitement drowned out the shout.

Mitzie snapped at the boy, and he started to chase her. He had not gone ten feet when Jake landed on his back. It was his first scream that terminated Bessie's break for a smoke.

The girl who was instructing the chaotic volleyball

class was only one year out of college, so no one could really blame her for not being able to control the surge of energy generated by thirty pre-teen girls, who ran to see what was the cause of all the barking and screaming. As they ran, they inadvertently formed a noisy cordon around the already excited dogs.

One boy, the one who had been pitching, managed to grab tightly onto Mitzie, but he let go quickly when she bit his mouth and tore a one-inch gash in his lower lip. Then Beau had the pitcher by the leg and was shaking his head furiously, while Clum was snapping at the boy's hand.

No one saw when it happened, but a pool of blood began forming along the first-base line, where Jake had torn a ten-year-old girl's throat.

The male gym instructor reached down to free the boy, but he automatically withdrew his hand when it was gashed by a late arrival.

As the acting principal ran up to the melee, she brought the final total of participants to three adults, sixty-three students, and eight panicked dogs.

The gym instructor and the acting principal together managed to get the young pitcher loose from Beau and Clum, who turned and ran back toward the snarls and screams coming from the first-base area. There Jake and Duke were both on top the first boy who had grabbed at Mitzie.

Packs of dogs attacking humans perform in a strangely predictable way: They center their efforts on their original victim, unless they are threatened by someone else. Even an intruder into the attack is not pursued as long as he retreats Jake's Pack was no different; it concentrated on the small boy Jake had hit first, except for the brief flashes when they would snap at the adults who were trying to stop the onslaught.

The screaming schoolchildren pulled back in horror as the dogs kept hitting the downed boy. Only the adults kept reaching in and kicking at the gnashing jaws. Two of the children and the girls' gym teacher passed out, contributing further to the pandemonium.

No one noticed when Mitzie slipped out under the legs of the harried children. She ran across the playground and disappeared around the school building.

While the Pack was frenzied, an element of leadership still existed. When Jake realized that his bitch had escaped, he fought his way free and took off back in the direction from which they had come. The other dogs quickly ran off, some after Jake and others in different directions.

As quickly as it had begun, it was over.

On the playground, in the general vicinity of first base, lay two children seriously injured, the first baseman and the pitcher. Fifteen children had minor bites and punctures and the three adults had extensive cuts on their hands. The other children had been subjected to a mental trauma that would leave scars for the remainder of their lives.

The cause of it all, the mischievous beagle, Mitzie, had fled to the nearby shopping center and found a hiding place in back of the Ashram Bakery. She cowered there in confusion.

It was three o'clock by the time reason had been able to reach out over the town and calm runaway emotions.

The most severely injured children had been whisked off to the Seibel Clinic.

In the school gymnasium, teachers, doctors, and parents established a first aid station to treat minor bites.

By the time Sheriff Fowler and his on-duty deputies had arrived at the school, the dogs had vanished. After several confused attempts to interrogate the children, he had wisely called the Morgan County Hospital and requested the assistance of the chief of Child Psychiatry to aid in the interrogation, and to give counsel as to how best to reduce the trauma in the children. Greasy had recognized the fear-glazed look in the young eyes and incessant chatter as shock.

Meanwhile, he had dispatched several of his patrol units to search for and capture the animals. He had horrible visions of the school kids and teachers being subjected to the painful anti-rabies treatment if the dogs were not brought under rabies-control observation. The anti-rabies treatment involves fourteen daily injections into the tender stomach area. If there was any possibility of rabies infection, then a lot of Brown County kids would suffer a second severe trauma.

Dr. Clay Pierce walked out of the gym and approached Greasy with slumped shoulders.

"Rough?" Greasy asked.

"It's not as bad as it looked at first," Clay said. "I've got the teachers scrubbing the kids. They'll be okay. But I'm worried about that one boy who was so badly bitten; he must have had a couple of hundred wounds."

Just then, Mel Ryan drove up in the park pickup truck. He got out and walked over to the vet and sheriff. "Oscar sent me over to see if I could help," he said.

Greasy shook his head negatively. "We got it pretty well under control, Mel. We gotta get them damned dogs, though."

"Quite a mess, huh?"

"Just no sense to it," Greasy said. "I guess the dogs just went crazy."

"I'd bet there was a bitch in heat involved," Mel offered. "That's how most of the packs gather up—chasing a female."

"Could be. One of the kids did say it looked like they were all chasing one dog. That's why the first kid grabbed the mutt; he wanted to protect it from the rest of the dogs. But after that those damned dogs just went crazy. Christ knows what they'll do next."

Chapter Nineteen

Mary Ellen had ridden alone for over an hour back at the training ring at the Saddle Barn. Then, after rubbing her horse down and loading him into the trailer, she had driven up by Mel's office. But she saw that his truck was not there, so she drove out of the park and went to the stable where Arab was boarded. Then she went home. She felt listless during the drive. People on the streets in Columbus seemed to get in her way more than usual; traffic was slower than normal. She finally realized that she was worried about Mel, and she liked the feeling. She hoped to see him that night.

She arrived at her home and pulled into the garage. Sitting there, savoring the realization that she had found a friend worth liking, she thought about her day—and that brought the dog incident sharply into her mind.

She wondered why Duke had not come to greet her car as he always did. Unless he was not home? That possibility caused her to shiver. If he was not there, then he might be part of that Pack of dogs.

She walked out of the garage and went to the kennel. To her relief, Duke was there, sleeping peacefully with his head hanging languidly out of his doghouse. She noticed his bowl was empty and decided to re-fill it for him. He'd probably be hungry when he woke.

She halted for an instant as she saw a large lump in the far corner of the kennel. At first she thought it was a pile of straw that the gardener had piled there for Duke. But, as she studied the mound, she realized that it was a dog curled cozily sleeping.

The discovery was a shock because Duke had never allowed another dog into the kennel; that was his absolute territory, and not even an innocent little poodle was permitted across the threshold.

Mary Ellen approached the open gate and reached for the bowl that was sitting on the ground right inside. At first she did not believe her ears when she heard the snarl. Then, as she glanced up, she saw that Duke's guest had awakened and was staring at her while emitting a low, nasty-sounding growl.

In a second, she was really shocked—because Duke was offering a similar growl.

She made a slight move to extract the bowl, then froze as the two dogs began to growl louder.

Mary Ellen remembered that her father had been emphatic when he had explained that the kennel was Duke's very private domain and that he should be respected if he showed a strong inclination to defend his possessed area. But this was more of a reaction than she had ever seen from him. Then she remembered how he had struggled with her the night before.

She eased herself into a standing position and began backing away from the kennel. She had moved several feet when she realized she should have closed the gate, but the dogs were both pulling themselves up, and she had no intention of racing them to it.

"Now," she said softly. "Be nice."

But the dogs continued to growl. Duke's uninvited companion, she saw, was a large collie. It was a pretty dog, if one could look past the bared fangs and the angry-looking eyes.

The collie was taking short, unhurried steps toward the open gate. Duke was close behind. Mary Ellen continued to back slowly away from them, but she did not feel comfortable with only a twenty-foot buffer; she had seen Duke cover such a distance in three or four leaps.

"Duke! Sit!" she commanded.

The collie growled even louder.

She continued to back away until she reached the corner of the garage. She thought of dashing for her car, but quickly abandoned that idea. Both windows were rolled down, and she dreaded the thought of trying to get them closed with two angry dogs trying to get inside.

With one fast glance back at the dogs, she bolted for the kitchen door about thirty feet away. Out of the corner of her eye, she had seen the dogs abandon their slow, menacing walk and break into a determined run.

Her feet had carried her five or ten running steps before she suddenly remembered that the cook and the gardener had been given the day off. Supposing the kitchen door was locked? "Oh, God, please let it be open," she prayed.

She swore, as she thought about it later, that she had felt the breath of either Duke or the collie as her hand hit the brass doorknob. Actually, the dogs were a good ten feet in back of her when she felt the door begin to swing in; the cook had left it unlocked.

The collie, unfamiliar with the gravel-covered driveway, slipped just as it was about to launch a final leap, giving Mary Ellen the brief instant she needed to slam the door shut. She stood there, breathing heavily, and tears began welling up in her eyes.

She quickly tried to remember exactly what Duke had done. She knew the collie had come after her, but was Duke's reaction directed against the other dog, or had he been acting in concert against Mary Ellen? She was just not sure.

She looked out the window and saw the two of them moving around the corner of the house. Suddenly she thought about the rest of the house; Duke knew it as well as she did, and if there were any other doors open he would know how to get in.

The swimming pool!

The sliding French doors leading from the living room to the pool area were never closed, except at night and during the summer when the central air conditioning was on. Duke would know about that.

As she ran into the living room, she could see the two dogs there at the screen door, pawing at the flimsy material. Duke had ripped it often before, and the collie was right there; if she got the idea they could be inside in seconds.

Mary Ellen bumped into an end table and a porcelain lamp fell, crashing loudly to the floor. The sharp noise caused Duke to shy for an instant, but the collie gave a strong leap forward and burst through the screen, snarling.

Duke barked angrily as he followed the collie through.

Mary Ellen ran backward for a couple of steps just as Duke crashed into the collie. The two dogs tumbled into a furious ball, and a floor lamp toppled to the floor with a crash. Without thinking, Mary Ellen screamed, "Duke!" and reached down to grab at his collar.

A searing pain at her wrist made her jump back. One of the dogs had bitten her.

Though it was difficult to see in the confusion, Duke seemed to have the collie down on the floor with his teeth at her throat. Mary Ellen backed away, moving behind the couch. She saw a heavy silver water pitcher and grabbed it up. It was full, so she threw the water at the thrashing dogs.

Startled, Duke jumped clear of the collie and snapped his head instantly in Mary Ellen's direction. The collie was on its feet again, too, and looking at her. She made a dash for her father's library.

As she ran in back of the U-shaped arrangement of couches and chairs, the collie made a leap calculated to intercept Mary Ellen's path. The dog cleared one sofa but landed awkwardly on the coffee table and lost its momentum and direction.

Duke was swerving between the end of the sofa and the fireplace, and he collided with the other dog. Fierce snarling and growling followed, and Mary Ellen paused a moment. The collie was on the floor again and heading around the couch, with Duke close behind. She grabbed a cushion and ran again for the library door, which seemed miles away. Just as she reached it, the collie's jaws caught her Levis at the ankle. The cloth ripped as she jerked her leg away. As soon as she recovered her balance she kicked out. The blow landed in the collie's throat; the dog howled in pain. She slammed the door of the library shut just as the collie banged into it.

She stood there, breathing heavily and trying to organize her thoughts. There was an ominous sniffing outside the door, and a deep-throated growling.

As soon as she had herself under control she went to the phone and tried to reach her father, but he was golfing and not available at the club. Her mother, as usual, was off shopping in Indianapolis. She thought of calling the police, but she feared that Duke might be taken into custody and even put to sleep.

She dialed the number for Brown County State Park.

In less than a minute, Mel was on the line. "How'd you know I'd be in the office?" he asked. His voice sounded tired.

Her voice was weak and frightened. "Mel, I've got a . . . I hate to bother you . . ." She was stammering, and her hands were shaking.

"Hey, hold on a second," Mel said. "What's up?" He waited for a response, but all he could hear was rapid breathing. "Mary Ellen? What is it?"

Her voice seemed to steady, but she was nearly whispering. "Just a second, Mel. Let me get my breath."

Just then Oscar Korteen burst into Mel's office, shouting. "I'll be Goddamned if I like this," he yelled at Mel. "You know what they're gonna do?"

Mel raised his hand and said into the phone, "Hold on a second, Mary Ellen. I'll be right back." He pushed the hold button.

Mary Ellen cried, "Mel! I need help!" But the line was a silent electronic void. She put the telephone down on the desk and walked cautiously toward the door. She put her ear to it and listened intently. At first she heard nothing. Then she tried again, lower down. This time she heard the sound of two dogs sniffing.

"Duke," she said softly. "Good boy, Duke." The sniffing stopped.

She waited a minute or two and heard nothing. She put her hand on the knob and started to turn it, intending to open the door a crack to see what they were doing.

Suddenly there was a snarling and howling louder

than any she had ever heard before, and something crashed against the upper panel. The door held, but she saw that the panel was split. There was another crash, and the split widened, and the noise turned into a scream.

And, as she slid into the blackness of unconsciousness, she realized for a moment that the screaming had been her own.

"What's wrong with that?" Mel asked.

"There's plenty wrong," Oscar said angrily. "Who the hell is going to explain it to the director? Tell me that. Greasy Fowler doesn't have to go up and fight for money. Shit! Why in my park?"

Mel did not know what to say, because he did not really see anything wrong with a meeting at the lodge. If things were as bad as it seemed, then there had to be a meeting someplace. The lodge was convenient.

"Lemme tell you one thing, boy." Oscar's voice was tense. "You go sucking in with that Greasy Fowler and you'll find yourself out of a job."

Mel sat there stunned. Then he said, "Don't be silly, Oscar."

They talked for a few more minutes, and Oscar began to calm down. He admitted that he was upset, and as he left he said, "I want you at that meeting, Mel. In fact, consider that an order."

"I'll be there," Mel replied.

He sat there for a few seconds, then remembered that Mary Ellen was waiting on the phone. He pushed the flashing button and said, "Hello. I'm sorry."

There was no reply.

Mel listened for a couple of seconds, then repeated her name. Nothing. He was ready to hang up and call her back when he heard a noise. He listened more intently. For the first instant or two he did not think he was hearing correctly, but it was there. The sound was weak, as if it was far-off, but it was there—dogs growling.

He yelled her name into the phone, but there was nothing on the other end except the faint sound of growls.

Mel slammed the phone down and jumped from his chair. He moved rapidly out of his office, poked his head into Oscar's, and said, "I've got to go out. I'll be back in a little while." Oscar started to say something about the meeting at the lodge, but Mel didn't wait to hear.

Mel made the drive to Mary Ellen's by breaking every speed limit on the route. As he pulled into the driveway he glanced at the kennel and saw the gate was open, but there was no sign of Duke. The only car in the parking area was Mary Ellen's so he assumed she was alone. He skidded to a halt, spraying gravel.

After a couple of insistent presses of the back doorbell, he tried the door. He moved quickly through the house and came into the living room. It was a mess.

He called her name and heard a faint reply coming from behind the library door. He went to it, opened it forcefully, and saw Mary Ellen sitting there in an overstuffed leather chair. Her face was white.

She offered him a weak smile. "I guess I passed out."

He went to her and felt her forehead. It was cold and clammy to the touch, but her eyes were steady so he figured she was on the recovery side of shock. He went to the bar and poured a shot glass full of brandy, carried it back to her, and ordered, "Drink that, we'll talk when you're through."

She drank, then stood and went to the window. "Are they still out there?" she asked.

"Who?"

"The dogs. Duke and a collie I've never seen before."

"I didn't see them."

She started crying then, and it was several minutes before she was in any condition to talk again.

Chapter Twenty

By nightfall, Duke and Vickie had crossed Indiana Highway 135 and penetrated the boundary of the park. The trouble at the house was over and all but forgotten.

During the afternoon attack on the children at the school, confusion had broken up the Pack, and it was several hours before they were able to gather together again.

Jake had been the first to leave because he had sensed Mitzie was gone. Beau and Clum had easily fought their way out of the mess of screaming children and violent adults. Aside from nearly getting run over by a car, their trek back to the base had been uneventful, except that the two new dogs had followed. They all arrived at the cave and settled in for a companionable rest.

Duke and the collie arrived at the cave shortly after dark. The Pack was assembled, including the two new dogs.

And Jake seemed to be waiting for something to happen. . . .

Mel switched off the headlights as he parked in back of the lodge. Beside him Mary Ellen was still wiping at her red-rimmed eyes. Once again he told her, "I'm sure Duke was only protecting you, Mary Ellen." He still did not know for sure what had really happened at her house, but if it would calm her, then it was worth it to continue the lie.

He had stayed there as long as he could, but time had run short and he had to be back at the park for the meeting. Mary Ellen's parents were not home, and he knew he could not leave her alone. The dogs might return, and even if they did not, she was in no mental condition to stay by herself.

On the ride back to Brown County State Park he realized that she had not heard of the dog attack on the school playground. Suddenly the thought of taking her to the meeting seemed to him an awfully bad idea. He stopped the car and explained the purpose of the meeting.

"Was Duke with them?" she asked fearfully.

"I don't know. I really don't. There was a Doberman. Probably the same one that came at me earlier today. But I didn't ask about the other dogs; there was too much confusion at the school. I'm sure we'll hear at the meeting."

In strained tones, she said, "I don't want to know." She paused, then in a small voice told him, "You say they—mauled a little boy?" She began to cry again.

"Look," he said. "I hope Duke wasn't with that Pack, but if he was, that's no reason for you to hate him or deny it. It's as natural as eating for a dog to chase and attack; it's up to humans to see that they don't have the chance to maul children. But if it upsets you this much, maybe I'd better take you back home."

She leaned over and put her head on his shoulder. He could feel tears coming through his shirt. Quietly she said, "No, go on. I'll be all right."

When they walked into the lodge, the meeting had already begun.

Sheriff Fowler had already spoken, and as Mel and Mary Ellen found their seats, Dr. Clay Pierce was speaking.

". . . and we have only nine days," the vet said. "After that we'll have to begin the rabies treatment for all of those kids, and that can be a painful process."

Mel looked around the room as the vet continued.

They were in the main lobby of the lodge, and all of the furniture had been turned toward the end of the room where the television set was. A table had been brought from the dining room. The sheriff and Oscar were seated there, and the vet was standing before it as he spoke. In the audience, Mel could see some people he knew, but many others he did not. There was the farmer, Luke Amoury; the acting principle, Bessie Carlton; and the camper who had lost the collie, Vickie. Mel guessed the other people were from the local newspapers or those of the surrounding communities. There were present several pairs of parents of the children who had been in the playground, a few employees from Brown County State Park, and all of Greasy Fowler's deputies.

"There isn't any question that rabies is a possibility," the vet explained. "I examined a dog today, and he was rabid. There is a good probability he was part of that pack."

"You bet your ass he was," Luke Amoury interrupted angrily.

Greasy Fowler raised his hand in a quieting gesture, "Okay, Luke. Let the vet finish. And let's watch the language."

Clay Pierce continued: "I'd like to capture those dogs and put them under a rabies-control condition so we can observe them during the incubation period. Maybe the dog I examined was the only rabid one. We can tell by keeping the others in quarantine for the nine days."

Luke Amoury stood up and looked around. It wasn't all that crowded around Luke, mostly because his clothes and boots were covered with chicken droppings. "I got a question," he said. "If it takes nine days, how come you found out that dog I shot last night had rabies?"

Clay explained that there were lab tests that he had been able to do on Blackie.

"Why can't you do them on those others, huh?"

"Because that dog was dead."

"You killed him, right?"

Coldly, Clay said, "I put the dog to sleep because it was the humane thing to do. He had been shot and was in misery. As you well know."

"Then you cut open his head and poked around, right?"

"That's the procedure, Mr. Amoury. We can check the brain cells of a dead animal to detect rabies." Clay was becoming impatient.

"Okay?" Luke said. "Then we can kill the others, right? And then you can examine them right away, without making those parents worry for nine whole days."

Sheriff Fowler stood up quickly. He indicated to Clay to remain standing. "Now, look, Luke. I'm sure all these folks know what happened, and we feel real bad.

You lost chickens and you got a right to be mad. But the first thing we gotta do is find them damned dogs. Let's work on that."

Greasy sat down.

Clay waited for a sign from the sheriff before he continued, then he said, "I have to agree with Sheriff Fowler. My main interest is to encourage you to be careful with the dogs. We don't need any more bite victims who might need treatment if we don't get the whole pack."

Clay Pierce probably could have done a lot of good for his career if he had sat down right then, but he didn't. He added, "Also, remember that those dogs did nothing wrong. They were only acting out of instinct. The problem——"

He was cut off with yells from several people: One of the parents complained loudly, the gym teacher held up his bandaged arms, and one deputy said "Bullshit!" loudly enough to be heard all over the room.

Greasy leaped up, motioning for Clay to sit down.

"Now, I'm sure the doc here didn't mean that the way it sounded. There's no reason to take offense by anybody. We just want to find out what we should do; that's why we're all here."

The hubbub continued.

Mel looked at Mary Ellen. Her face was pale.

The Pack had felt secure for a while, but an overt change had taken place in Jake.

They had regrouped in their hidden sanctuary and rested through the afternoon and into the shank of the evening. Occasionally they would wake, readjust their positions, look around, and then go back to sleep. But each time they noticed that Jake was always awake. Sometimes he was reclining, sometimes pacing restlessly, but he was constantly near the entrance.

As dusk had fallen, Duke and the collie arrived, and Jake firmly herded them back into the shelter with the other dogs. He returned to the cave entrance, and to his restless pacing.

There had been several things that caused Jake to act the way he did. There had been an unusual amount of activity around the lodge earlier, with cars driving up and strange voices coming through the night air. There were other things: The animals in the park, also aware of the noises at the lodge, had gone silent. Nature seems to have endowed dumb animals with an ability to sense danger, a foreboding that is a special tool for survival. It does not always work: A deer might sense danger and tense up as a hunter draws a bead, then go back to munching grass as the trigger is pulled.

Jake's warning came from some source unknown to man, but it was making him anxious.

The meeting in the lodge was nearly getting out of hand, and Greasy Fowler felt he was walking a tightrope. He was not about to take a firm position until the will of the people was expressed with some unanimity—he was not a three-term sheriff by accident.

"Quiet!" he yelled. It was a couple of minutes before he could be heard. "Now, all of this ain't doing us any good," he said. "We gotta decide what to do and how to do it. I know you folks whose kids got bit are here to help, and you all know how bad we feel. We'd like your help. But it ain't goin' to do no good if we just keep yelling." The room got quiet.

"The main thing," he went on, "is to get the dogs. We gotta do that. I'd like Mel Ryan to tell you what he found today."

Oscar made a move to protest but was cut off by a stern look from Greasy. As the sheriff sat down, he whispered to Oscar, "Now, you don't worry. We ain't gonna make your park look bad."

Mel stood up, looking uneasily at Oscar. "I don't know what I'm supposed to say," he began. "But I do think I know where the dogs are hiding. As he told the story, he included all he could remember about the terrain and location. "So," he finished, "I'd guess they have a base that is well-hidden, not from man but just because it would offer good protection from the elements. Don't get me wrong; if these dogs are feral, they will fight. The one sure mark of a feral dog is its viciousness. No petting or talking or offers of food are going to change that; he'll take a whack at anyone who threatens him."

"Then we gotta shoot 'em. Right?" Luke Amoury said.

"I didn't say that, sir." Mel looked confused, and Mary Ellen saw him looking uncomfortably in her direction.

Another controversy exploded, and Mary Ellen could not believe what she was hearing: Quite a few of the people at the meeting were beginning to argue in favor of slaughtering the dogs.

The vet, Clay Pierce, was on his feet and, after considerable effort, Greasy Fowler restored a reasonable calm. He motioned for the vet to speak.

Pierce said, "I was not asked to come here as anything other than a source of information on the health status of these dogs. But I live in this town, too, and I don't like what I'm hearing. I know a lot of you farmers go around shooting dogs, but there is no reason for all of us to get trigger happy. First, those dogs *might* be feral—but they also might be perfectly normal, home-loving, flea-scratching pets. There is no reason to slaughter them. I'll even go on record right now: I'll take any dogs that are not wanted and I'll keep them. I happen to feel that even a feral dog can be redomesticated, and I'm willing to try."

Luke Amoury stood up again, red-faced. "You think we're gonna go hightailing it out into that woods unarmed and run the chance of them dogs comin' at us in the middle of the night? You crazy? One of them dogs comes at *me,* I'm gonna start shootin'."

Oscar stood and yelled at Amoury. "I'm beginning to think you're the one who's crazy," he shouted. "And *nobody* is goin' off shooting in this park. I mean, nobody!"

The people at the meeting were beginning to choose up sides.

Jake decided to move.

It was not puppylike curiosity that brought him out of his cave; he was boss dog, and the Pack's welfare was his responsibility. That same responsibility had brought him out earlier in the day when he had smelled man approaching the territorial area. That time he had led Beau and Clum on the investigation; this time he would go alone.

The Pack began to follow as he started through the woods, but he growled at them. They followed as far as the clearing, but stopped there and waited.

A Doberman's gait is nearly unique in the dog world. He moves with a strong reach of his forward legs, and his rear legs seem like pistons as they deliver extraordinary driving power. He does not arch his back as he moves and his feet, when moving at a fast trot, track perfectly in line. Jake moved through the dark woods, using every inherited quality that generations of breeding had given him. He reached the lodge in minutes.

There was some activity at the back of the lodge, but that was at the rear doors of the kitchen and involved only the familiar removal of garbage. He moved in the shadows made by the utility lights shining down on the parking lot, and his catlike paws were silent as he walked onto the asphalt.

He sniffed.

There were noises coming from within the building, gruff sounds and angry words that reminded him of his former master at the service station.

He sniffed again.

Then he smelled it, a familiar odor that he did not like. He followed the scent and approached the source.

Three of the sheriff's patrol cars were there, and that was the odor he disliked—a faint residue of carbide, the smell of highly polished leather, the pungent remains of stale sandwiches. They all reminded Jake of when his former master had played at being a cop.

He backed off slowly with breath puffing silently out from his flews. Three steps back and he spun around and sped off back into the woods heading for the cave.

"I DID NOT!" Mel jumped to his feet and nearly knocked Mary Ellen off the chair beside him. "I did not suggest that we kill those dogs."

Greasy Fowler raised both hands, palms out in a quieting gesture. "Hey, Mel, take it easy," he said in a mild manner. "You just told us it was heavy going back in those woods, and you said the dogs was probably pretty well hid. You said they'd attack us, if they was—what'd you call it?—*fee-rial,* didn't you. I just assumed you was suggesting we shoot 'em."

Greasy had covered all of his bets. He'd tricked the vet into admitting that tranquilizer guns would be nearly impossible to use on a pack of dogs in the woods at night; he'd played to the parents of the injured children to get them to be vocal about getting rid of the "dog problem"; and he'd reviewed Mel's report and implied that Mel had suggested shooting the animals. Once Greasy had gotten the idea that the majority of people wanted something done immediately, he had taken the tack that shooting would eliminate the problem most easily. In fact, he had convinced himself that they had no other choice.

The sheriff looked toward the back of the room, where a man had just stood up and was waving for recognition. "You got somethin' to say, mister?" he called out.

"I've been quiet during all of this," the man began, "but I think I've got some right to talk. I'm Jeb Lewis from Chicago, and my dog is lost somewhere in that park. I don't see where you've got any right to talk about going out and shooting my dog."

There was a slight murmur in the room.

"Our Vickie is about as nice a dog as you can find. She's been really wonderful with our children. She's like another kid to us. Now I understand why some of you—" he pointed in the direction of Luke Amoury—"would demand such a thing. But our Vickie was locked up last night. She didn't get out until this morning, so she couldn't have killed any chickens or gone after the deer. From what the principal here has told us, it sounds like there was a collie at that school ground, but it could have been another collie, couldn't it? Vickie would never bite a child. My God, she's been with babies and children all of her life; she loves kids. So, I'll tell you right now. You go shooting at my dog and you have me to answer to."

The room was silent.

Mary Ellen stood up and moved close enough to whisper in Mel's ear, "Please tell them about Duke. It's the same as that man. Duke wouldn't attack children."

Mel turned and looked at her. He shook his head and whispered back, "We'll take care of that later. There's a nasty mood in this room and I don't like it."

She sat down.

"Well," Greasy said as he cleared his throat unnecessarily. "Well, now, Mr. Lewis, I want to thank you for your good words. None of us here wants to see your dog shot, and I'll tell you right now that I personally will guarantee that it will only be shot if it attacks. Okay?"

"What the hell are you talking about?" Jeb Lewis said angrily. "Who's to decide if Vickie is attacking? Will it be that crazy farmer over there? Or, one of your trigger-happy deputies? What right do any of you have to go around killing dogs?"

Greasy smiled inwardly but forced his face to seem stern; Vickie's owner had played right into his hands. The local people would not like some slicker from Chicago making fun of the farmer or the deputies. He looked back at Jeb Lewis and said, "Now don't go insultin' these good people. I gave you my word, so let's let that drop."

Greasy had a feeling for audiences; his years of constant campaigning had sharpened his grasp of such things. He waited and let them chat back and forth for a couple of minutes while he acted as if he was busy with some blank papers on the table in front of him.

During the lull, the problem of how to get them all to agree was solved for him. He saw one of his deputies waving to him and pointing to the telephone, and he left the front of the room and walked over. He picked up the phone, listened for a minute, and then hung it up. When he went back to the table at the front of the room, his face was grim.

"Folks!" he said loudly. "Please, I need your attention for a minute here!" The noise subsided a bit, but Greasy waited till it was quiet. Then he said, "It seems like while we was all arguin' here about whether to shoot those dogs or not, the doctors over at the hospital have been fightin' to save that little boy's life." He paused dramatically. "Well, they lost," he said slowly. "The boy died five minutes ago. And those dogs killed him."

There was a stunned silence for a few seconds, and then everyone was on his feet talking at once.

Mary Ellen grabbed Mel's arm. "Talk to them, Mel," she said. "Please. They're going to shoot those dogs, and Duke may be with them."

At that moment, Greasy clapped his hands for silence. When it was quiet, he said, "Now, I guess there ain't much doubt about what we ought to do. All we need to do now is get ourselves organized to get the job done."

Mel stood up. "Wait a minute, please," he said above the din. "We can't all just go out and start blasting away at everything that moves." He turned to Oscar. "Tell them, Oscar. We can't turn the park into a battleground."

Greasy's eyes flared slightly but he kept himself under control. He turned to Oscar. "Nobody said anything about shooting up the park, Oscar. We'll work it out so we don't have to do any shooting in your park. And will you please have your boy there shut up?"

Face burning, Mel turned to Oscar. "Listen, Oscar, they're talking about killing dogs," he said earnestly. "Don't you understand what that means? We can't let them go out and shoot up every dog in the area."

Without another word, Oscar left his seat behind the table and walked to Mel. In a hushed voice he said, "You come with me. I want to talk to you."

Perplexed, Mel reached down for Mary Ellen's hand. The two of them followed Oscar out into the cool evening.

They all stood there in the darkness on the lawn in front of the Abe Martin Lodge. Light spilled out of the windows, and the sky to the north glowed from Indianapolis thirty miles away.

Oscar, as usual, bought time to think by lighting a cigarette. His face showed a red tinge as he pulled a deep puff before replying. When he did speak, his voice was calm and steady, with an almost fatherly tone. "Look, Mel, I've taken a real liking to you, you know that?"

Mel nodded.

"Okay," Oscar went on. "Then you know I want to keep you on at the park. But, Mel, I can't have you messin' up my job. Now, Greasy has promised that he has a plan that won't even involve the park, and that's my big concern."

Mel shook his head in disbelief. "You say 'plan' and 'concern' and 'involve,' but that sonofabitch is in there talking about shooting dogs. Don't you realize that?"

"I do, Mel. Hell, I'd bet old Greasy had that plan all thought out before the meeting even started. As sheriff, Greasy don't like nothin' bad hangin' around too long. That's the way he is."

"We can *try* to change him. There's no reason to kill those dogs."

"No? Tell that to the parents of that dead boy."

Mel's shoulders slumped in frustration.

Mary Ellen spoke for the first time. "Mr. Korteen, I don't know why you're so set on following the sheriff, and I probably don't have any right to ask. But there is a good possibility that my dog is out there with that pack. My dog is a good dog, Mr. Korteen, really. Just like that man from Chicago, I know my Duke is good. I've had him since he was a puppy." She broke into tears. "I don't want Duke shot. *Please.* You *have* to stop them."

Oscar looked away.

Mel moved to Mary Ellen and put his arm around her shoulders. "Easy, take it easy," he said.

She pulled away from Mel. "No, I won't take it easy," she said. "You men make me sick with your guns and killing." She walked a few feet away and stopped, glaring out into the darkness.

Oscar moved closer to Mel and spoke in a low voice.

"You'd better get that kid out of here, Mel. She might know her dog and you might know everything there is to know about dogs, but the time for discussion is past." His eyes narrowed as he looked intently at Mel. "I want you back in there. I want you to help out on this, and to see that nobody does any shooting in the park. Is that clear?"

Mel heard himself saying, "Okay, but one thing. I want your promise that if there's *any* chance to save *any* of the dogs, you will support me."

Oscar thought for several seconds before replying. "Okay. But I don't think there's much I can do now." He walked quickly back up onto the stone porch and into the lodge.

Mel walked over and put his arm around Mary Ellen again. She was crying.

"He said he'd help us look out for Duke," Mel told her.

"Oh, sure," she said.

He held her a little tighter, and she turned to look up at him. She tried to smile. "I know *you'll* try to save him. I know you will." She was quiet for a moment, then she began to cry again, but it was a milder, less angry cry. Softly she said, "Poor Duke . . . oh, God . . . poor old Duke."

Chapter Twenty-one

At that very instant "poor Duke" was flat on his back, his four feet up in the air and his mouth full of Beau's left ear.

A not too well-known Irish philosopher named John Fox once observed that "Winter is a dog's summer," and anyone who owns a dog knows how frisky and playful he will become when it is cold. It was beginning to get cold that night and the dogs needed activity. While Vickie and Clum were splashing down in the coolness of the stream, Beau and Duke played by rolling in the damp moss and biting each other's ears, necks, and legs.

Beau emitted a slight whimper and Duke released the ear from his mouth. As if in contrition, he rolled over on his side and left himself vulnerable. Beau shook his head, flapping his ears as if to shake away the residual pain of Duke's bite, and he lunged down at the boxer's throat. Duke took the onslaught for a second as Beau mouthed a concerted mock attack on the tight skin of the boxer's neck. Then, when the mouthing became too vigorous, he placed all four paws under Beau's brisket and pushed. Beau was catapulted into the air and landed with a thump on the ground. Before Beau could regain his breath, Duke was over him biting noisily at Beau's mouth.

The game stopped instantly as they heard Jake running through the woods on his way back to the cave.

Duke emitted his familiar *grruuufff* as Jake made one final, strong leap through the underbrush and landed in the clearing. The mood of the group changed suddenly as the powerful Doberman stood there. His stance was not playful.

Beau's ears folded back along his head, Duke's chest expanded, Clum's tail erected and curled forward. All the dogs were silent, as alert as Jake . A crisis seemed imminent.

The atmosphere in the lodge was thick with tension as Mel walked back in.

He had put Mary Ellen in the park pickup truck and told her to wait for him. She had still been crying when he left her there.

As he looked at the people in the lobby of the lodge, he was glad Mary Ellen was not there; he did not like what he saw and felt.

He walked over to the vet. "What's up, Clay?"

Pierce shrugged his shoulders. "I don't see any way to stop them. I wish to hell Fowler would listen to reason. I hate to see those dogs slaughtered."

Mel said hopefully, "Maybe they'll get away. I think that Doberman is one smart dog."

"It could happen" Clay replied. "Any dog that has been able to survive more than a month in the wild becomes a pretty resourceful animal. If we go crashing around out in the woods, they might shy and take off. That in itself could break up the Pack. I hope so."

"What have they cooked up, Doc?"

The vet shook his head, "I really don't know. I'm

sure they don't want any of my advice." He motioned toward the group around the table. "Go ask them to tell you. While you're there, you might ask them if they'd even consider letting you and me use tranquilizer guns. It could save some of the dogs, maybe."

Mel walked to the front of the room where Greasy Fowler was going over the review of his plan.

The men gathered around the large table were in a festive mood. Maps were being studied, premature boasts were being made, and everyone moved with anticipation. They were lacking a hunt master with red jacket and brass hounds' horn; there was no mulled wine to be drunk while standing beside a Virginia-bred jumper who would take them to watch a fox be torn to pieces. These men were dressed in their daily clothes, they sipped lukewarm coffee, and they would use pickup trucks, autos, and police cars. But the spirit of the hunt was there.

The plan, as Mel heard it, was simple enough. The hunters would place themselves strategically along State Road 135, which runs near the eastern boundary of Brown County State Park. This was an obvious concession made by the sheriff to Oscar—the road was off park property. The hunters would use a couple of newborn lambs as bait, and they would wait in the dark for the dogs to appear. (The idea of using the lambs was from Luke Amoury, who had pontificated on the inability of dogs to resist the smell of a freshly dropped lamb.) They had broken into two groups: The women had gone home to look after their dog-bitten children, and the men had gone to get their guns. (They had asked Sheriff Fowler for the loan of guns, but Greasy did not think he should hand out official police guns to ordinary citizens.) They would assemble out at the highway.

Greasy gathered the remaining men around him. "Remember," he said, "we're after killer dogs. I think our community will greatly thank us tomorrow if we eliminate these killers tonight." He stood up. "Let's go."

Mel held back until Greasy was alone at the table.

"I'd like to ask a favor, Sheriff."

Greasy busied himself with folding up maps and papers.

"The vet and I want to take out tranquilizer guns. Maybe we can capture a couple."

Greasy exploded. "Ain't you got no sense? Them damned dogs is killers, Mel. Use yer head."

Mel swallowed hard. "Look, Sheriff, we'd like to try. Okay?"

"Shit!" Greasy said in a tone of disgust. He threw the papers into his briefcase and slammed it shut. He started to walk away but stopped. "Okay, you do that if you want. But I'll tell you one thing. If any dog gets past you alive, you'd better believe my deputies will be givin' you tickets for drivin' and summonses for spittin' for the rest of your life in Brown County. And that's no bullshit."

The sheriff stormed out of the lodge, throwing an angry glance in the direction of the vet.

Clay walked over to Mel and said, "He's a mean one."

Mel said, "You heard him, though. We can use the dart guns. You have them with you?"

The vet nodded. "Are you sure you want to? He'll make life hell if we goof up."

Mel thought for a moment before replying. He had

come to like the life in Brown County. Finally, he said, "Sure, we've got to give it a try." Then, looking at the activities outside through the window, he added, "I don't think Greasy's all wrong. He's got a rotten problem and he's doing the best he knows how. It's tough when a kid gets killed. If people would only think before they let their dogs run loose. . ." He started toward the door. "But right now let's see if we can save at least some of them."

Outside, Oscar was standing by the pickup truck, talking to Mary Ellen.

As Mel walked toward them, he watched the patrol cars and pickups as they screeched tires in their haste to get to the rendezvous. He told Oscar that he and Dr. Pierce were going to try to use the tranquilizers if possible.

"Yeah," Oscar said. "So I heard." He reached into the back of the pickup and lifted out a small bundle. "Here," he said, "I want you to have this, too. Just in case. . ."

He handed it to Mel. It was a Colt Diamondback .38 Special revolver in a leather holster. Oscar showed that he was carrying one also. "I always keep them in the office. I never knew why until tonight."

Mel felt slightly foolish putting the gun on, but the memory of that attacking Doberman in the woods was still fresh in his mind.

Dr. Pierce walked over with his tranquilizer gun in hand.

"Here, Mel. Use it well. I've loaded it for a seventy-pound dog. If he's much smaller, don't shoot him because the dose will kill." Mel nodded as the vet went on. "And, if he's much bigger, all I can suggest is that you run. The amount would make him sluggish but still dangerous. Okay?"

Mel laughed. "That's terrific. Will I be able to outrun him in the woods at night?"

The vet shrugged and went back to his own car.

Mel joined Mary Ellen in the truck and followed the disappearing taillights. They were nearing the exit of the park before she spoke. "We will be able to save Duke won't we?" she said in a soft, almost childish voice.

Mel concentrated as he drove out through the covered bridge. "I sure hope so, Mary Ellen. I'll do all I can. I want to save all the dogs, if possible."

Roaming packs of wild dogs do not have to concern themselves with all that much when they decide to move. They have no possessions to pack or bills to pay. They simply move on.

But there is a transition that must take place. Normally, the packs have a central base where they bed down. In the summers it is generally down low, near cooling mud flats and fresh streams that supply water. In the colder periods they tend to get up to higher lands, where tree cover is less and the warm sunlight is more available.

No matter where the packs center their activities, they will range individually or together over a broad area, generally two or three miles in any direction. The trips can be according to an established pattern, or they may be random. Sometimes they can last for a day or two.

But this move by the Pack was different. It was being launched because Jake had commanded it.

It was not all that easy to get all of the dogs going in the same direction at the same time. Beau and Clum were no problem; they had fallen into the pack structure well and accepted Jake's lead. The two new arrivals and the collie knew that Jake was boss dog, but they could not yet comprehend what a pack leader was, so they dawdled. Jake kept having to come back to get them moving. But Duke was presenting a different problem. He sensed that a major change was in the offing, and that had made him think of home, and of the food bowl Mary Ellen would fill. He had not decided if he should go home to Mary Ellen or stay with the Pack.

In the end, Jake was able to herd Vickie and the two new arrivals along in the direction Beau and Clum were traveling. Duke followed along behind, moving tentatively and avoiding contact with Jake.

They headed east from their cave, toward the boundary of the park, in a direction that would take them to State Road 135.

Chapter Twenty-two

Route 135 runs south from Route 46 through Gnaw Bone Camp to Stonehead. It is surfaced with gravel mixed in tar and is similar to thousands of such back roads throughout the country. It wanders carelessly, is just about wide enough for two cars, and the shoulders roll off either into ditches or flat, grassy pastures.

The location selected for the hunt was well picked. It was on a line running slightly north by northeast from the fire tower in Brown County State Park. The perimeter, as Greasy referred to his line of hunters, was spread out over a mile of Route 135, with patrol cars or pickup trucks spaced about an eighth of a mile apart.

Mel had been assigned a position opposite a field just north of the dentist's house and office. He sat there in the dark with Mary Ellen poised silently beside him.

The next post was a patrol car with the deputy sheriff and one of the parents. Then came the gym teacher with another deputy. And so it went on down the line, which was supposed to function as a catch net no matter where the dogs appeared.

At the center of the line was Greasy Fowler along with the farmer, Luke Amoury.

All of the vehicles involved in the operation were parked crosswise in the road, pointing their unlit headlights at the wooded area from which the dogs would probably emerge. The two newborn lambs had been tethered in a gully about two feet deep about thirty feet from the road. One was tied in front of the post manned by Oscar and a park employee; the other was in front of Greasy and Luke Amoury. The night was quiet except for the occasional bleats from the lambs.

"Think they'll come?" Greasy asked softly. There had been no sign of movement in the woods for close to half an hour.

"Yep," Luke replied. "If they're around, they'll come. A dog can't keep hisself from the smell of a new lamb." But Amoury was pacing around. He seemed impatient about something. Finally he said, "I'd shoot every one of them friggin' dogs if you'd just give me a chance."

They had been leaning against the front fenders of Greasy's patrol car, the sheriff on the driver's side, so he could flip on the headlights quickly, and Luke on the passenger's side. Greasy felt the car move as if Luke had stopped leaning. The sheriff squinted his eyes to see better in the dark. He spotted Luke moving toward the gully.

"Where the hell you going?" Greasy whispered loudly.

"I'll be back," was the hushed reply.

In the darkness of the night, Greasy could barely make out Luke's form. There was a small amount of ambient light from the town, but the moon had not yet risen and seeing was difficult. The sheriff listened intently and heard Luke's boots moving through the brush. Then there was silence.

Suddenly, shockingly, a terrible bleating erupted into the quiet night. The sound lessened in a moment, but continued on at a lower pitch. It was an ugly sound and made Greasy uncomfortable.

Luke came back to the car and resumed his leaning position on the right fender.

In an urgent whisper Greasy asked, "What the hell was that noise?"

"Jest fixed the bait," Luke said quietly.

Greasy pulled his flashlight from his hip pocket, walked through the underbrush, and beamed down into the gully. He switched it off as soon as he saw what Luke had done: The lamb was standing there, bleating piteously. Its nose had been slit open with Luke's knife, and blood was spurting onto the ground.

Greasy returned to the car. "You lousy bastard," he said angrily. "That was a rotten thing to do."

Casually, Luke said, "Yeah? Well, it'll get those dogs here fast. You better get ready now."

The Pack had moved slowly at first.

Jake was at the point of the grouping. Duke had moved into a position slightly off to the left of the Doberman and a few feet behind.

Vickie and the two new dogs bunched up in back of Jake and Duke, while Clum and Beau moved loosely out on the left and right flanks.

It had begun in a walk, but by the time they crossed a ditch called Pendelton Creek, about three quarters of a mile from Route 135, Jake broke into an easy lope.

When the bleating of the injured lamb reached the dogs, they stopped for a moment and listened. There is a sound in nature that animals all react to, a mysterious combination of frequencies that signal an animal in trouble. Almost automatically, the Pack began to move in that direction. They moved at a cautious trot.

By the time they were a half a mile from the bleating lamb, the scent of blood reached their noses. They increased their pace. Clum let out a hunting bark and the collie Vickie, echoed him, but a growl from Jake silenced the pair. After that, all that could be heard was the pounding of their paws on the floor of the forest.

Jake began puffing air out of his lips and he surged forward. At their steady trot, they were just under four minutes from the lambs.

Mary Ellen sat up with a jolt.

"I heard barking," she said tensely.

"I heard it, too. Take it easy; it could just be a farm dog yapping at all of the activity."

He reached over to take her hand. In the red glow of light from the two-way radio, Mel could see her strained smile.

She said, "I think we've been wrong—about Duke, I mean. When we get him back, things will be different. You wait and see." The tension began to creep back into her voice, "When I tell my father—"

Mel cut her off with pressure on her hand. "Wait," he said. He reached down and picked up the dart gun. "Stay put. I'm just going to get out for a minute."

As he opened the door, the overhead light flooded the cab of the truck. He looked at her. She had been wiping at her eyes and they were heavily bloodshot.

"You're a mess," he said quietly, smiling.

She nodded and tried to smile, but tears came back into her eyes. He closed the door and the light went off. then he stood there for a moment, breathing the cool night air, and realizing that he had begun to feel very close to the girl, even though they had only met the day before. It seemed he had known her much longer than that.

Headlights came down the road, and Mel spun around. The sheriff had posted a deputy up there to stop traffic, so he assumed this was the deputy. Then he saw that it was a large camper, and he remembered seeing it before. It belonged to the man from Chicago who had lost the collie.

Mel walked into the road and waved for the vehicle to stop.

"Hi, Mr. Lewis. Can I help you?"

The man leaned out of the driver's window and spoke. "You're from the park, aren't you?"

"Yes, sir, I talked to you today."

The man put the camper into neutral and leaned back out. He said, "I came down here to talk to the sheriff, but you'll do. You tell him for me, okay?"

"Tell him what?"

The man was angry. "We heard what you are doing down here; everybody in the park's talking about it. Well, we're leaving. I want you to tell the sheriff we think this whole place stinks. We're getting out of here before you go shooting our dog. We don't want our kids to know about that."

"We hope we don't——"

"And when I get back to Chicago, this is going to be in the papers. Now, you understand that?"

Mel wanted to say something, but he did not know what to say.

Jeb Lewis put the camper in reverse and started backing up to the dentist's driveway to turn around.

In back of him, Mary Ellen said, "That poor man. What's he going to tell his daughters?"

For the second time in less than a minute, Mel could think of nothing to say. He walked over and touched Mary Ellen's arm.

"It will be okay, won't it, Mel?"

At least he could still lie. He said, "Sure."

The dogs were moving at a full run now, and Jake was just over five hundred feet from the lamb when he caught the first faint scent of man.

He had pulled considerably ahead of Vickie and the two new dogs, but Duke, Beau, and Clum were still holding their own.

The other animals in the dense woods had long since stopped moving. Squirrels had tightened into their nests, rabbits had burrowed deeper, and even the nocturnal opossom and skunk had interrupted their foraging. The bleating lamb had been a cause for alarm; the running dogs were a cause for fear.

When Jake first smelled man, he dropped his hindquarters and slid to a halt. Duke and the other dogs stopped, too, and milled around the leader in confusion. The Doberman was tense, ears pricked, head erect, and stubby, docked tail rigid and unmoving. Beau, too, was frozen. His forelegs were stiffly extended slightly in front of him, his shoulders were slightly hunched, and his head reached anxiously forward so that the whole top of his body formed a straight line from his nose and down his back to his tail. His hackles were not up as Jake's were, but he, too, smelled man.

Jake moved forward, but in a slow and cautious manner. He was silent now, and his body slithered around branches that might snap. He walked carefully, using the podalic hairs between his toes to place his feet so as not to crackle leaves on the ground.

By that time, all of the dogs had picked up the human scent, but they waited where they were, unsure of what Jake was doing. For most of them, the man smell was familiar, and no cause for alarm.

Through the night came another mournful bleat from the lamb. The blood odor permeated the air and overpowered the human scent.

Suddenly, disorder came to the Pack.

Vickie leaped forward, and Beau moved with her instinctively. When Clum suddenly crashed ahead, the two new dogs followed.

Jake and Duke were left behind but only for an instant; they, too, succumbed to the urge for attack, and they broke into a full run to catch up to the other dogs.

In seconds they had all left the woods and were on a flat stretch of meadow, bounding through tall, wild grass.

Jake pumped his legs hard as he fought to regain his position at the head of the Pack. Duke was running strongly at Jake's shoulder. Then, as they plunged through the grass and began overtaking Clum, the pointer let out an excited loud bark. Vickie joined in and then Beau contributed to the chorus of the chase.

The blood scent grew stronger, and the pack ran in a tight group toward the source, abandoning their attack formation since the prey was not moving.

They were only about twenty feet from the tethered lamb when blinding lights suddenly burst from the darkness. Jake began to slow his run, intending to veer away from the lights. Before he could change his direction there was a thunderous blast right in front of him, and Beau flipped up into the air. Blood splattered down like a rain shower. The 30-06 bullet from Greasy Fowler's rifle had caught Beau just below the left shoulder, and he was dead before he hit the ground.

Jake saw Duke making a swing to the left, veering for the possible protection of the ditch beside the road. The collie was cutting across Duke's line of run, and she caught the next shot well below the neck in the front. Had she not dodged in front of him, the bullet would have caught Duke.

Jake bumped into Clum as he turned to follow Duke toward the ditch, and the pointer fell in beside, matching Jake's powerful run. Ahead of them, Duke disappeared from sight as he splashed down into the trickle of water running in the gully. The force of his impact stunned him for a second, and as he shook his head to clear the confusion, he looked up to see Jake soaring gracefully through the air over the ditch. Clum was an instant behind.

Duke scrambled quickly up the side to the ditch, clawing at the soft dirt and grass in an effort to follow Jake.

Jake was angling away from the headlights; then, off to the left, more headlights burst into the night. There was a loud yell from that direction, followed by a shot. Jake was forced to swerve back into the first blinding lights. Clum had not changed direction, and had now taken over the lead.

They were just at the edge of the road when another shot cracked and Clum's left hind leg flew out from under him. Jake slowed his pace fractionally and glanced back; Clum was tumbling and sprawling in the rough gravel, yelping in agony.

Duke had nearly caught up to them when the shot hit Clum. In the next instant, he saw Jake vanish into the darkness in back of the lights. Duke dove for the blackness, but he was still twenty feet away.

The same headlights that had been so blinding when the sheriff had first flicked them on now bounced back light from the road's surface. Jake could see a man in coveralls yelling and jerking at his gun.

"Damned gun's jammed!" Luke Amoury shouted. He ran to intercept Jake, still trying to work the bolt.

As Jake launched himself at the farmer's throat, a voice from the other side of the car screamed, "Look out!"

Greasy Fowler was trying his best to get his rifle up over the hood of the car, but by the time he got it into firing position, the Doberman had plowed into Luke and the dog's strong mouth slammed shut on Luke's throat. There was not even a scream as Luke fell backwards with the dog on top of him.

Blood dripped from his jaws as Jake spun his head around. Clum lay jerking and twitching in the flood of the headlights, a pool of blood spreading on the black road. Shadows rippled across Clum's inert body as Greasy Fowler ran around in front of his car, his rifle half-raised into the shooting position. Suddenly, from nowhere, Duke slammed into the sheriff, clamping down hard on the left arm and pulling it away from the stock. The gun crashed to the ground.

Greasy kicked out at the boxer, and Duke's feet grabbed at the ground, backpedaling frantically.

Greasy grabbed quickly for his service revolver. He had unsnapped the retaining strap when suddenly there was a searing pain in his leg.

Drawing on some tiny wellspring of strength, Clum had fought through the haze of darkness that was beginning to close down over his eyes. He had seen Duke fighting with the man, and he had dragged himself painfully across the road. He had reached the fight just as Duke fled, and had barely the strength left to close his jaws on the man's leg.

Greasy yanked his revolver out, flipped it around, and hit Clum in the head. The dog fell unconscious before he died, but the time the sheriff had used gave Jake and Duke the chance to speed across the road and into the darkness.

When the headlights had come on farther down the road, Mel had felt his pulse quicken. But, when the firing started in the direction of the sheriff's car, he felt a sick sensation in his stomach.

Mary Ellen called to him, but he did not answer. He shifted the tranquilizer dart gun from his right hand and slipped it under his arm, ready to bring it up into a firing position. Then he reached down and pulled the revolver out of its holster. He checked the safety and put the gun back.

"Mel!" Mary Ellen called again. "Mel, what's happening?"

He walked back to her side of the truck and said, "I don't know. Take it easy."

The shooting down the road lasted for less than a minute but it was too far away to see what was going on. He did see that most of the other vehicles had switched on their headlights, but he decided it would be best to keep his off.

In the distance he could hear yelling.

"Did'ja git one, Greasy?"

Mel winced.

"Hey, Sheriff. I got one o'them critters."

Mel looked at Mary Ellen, whose eyes were wide in her pale face.

"It doesn't prove *anything*, Mary Ellen," he said. "There is no way to know if Duke was even there." He could hear engines starting up along the road, and some of the lights were beginning to move. It suddenly hit him that he had not heard the sheriff answer the yells. He turned to the girl. "We'd better go down there, Mary Ellen," he said.

He went around the truck to the driver's side, and he was reaching for the door handle when Mary Ellen whispered loudly, "Mel! Look!"

Down the road, the lights from all of the other cars and trucks were casting a glow, and nearer by there was a faint silhouette of something moving toward them. Mel squinted into the darkness, and the image came into focus.

It was a dog walking along the side of the road. A large dog.

"Stay still," he ordered Mary Ellen.

He stood frozen except for a slight movement as he slowly brought the dart gun up into the firing position.

The dog kept coming.

Mel whispered, "Lean over and turn on the lights."

He could hear the rustle of Mary Ellen's dress as she slid across, reaching for the switch.

"Is it Duke?" she pleaded in a hushed voice.

"I don't know," he said. "Turn them on."

Light flooded the road.

The Doberman, startled by the flare of light, stopped in his tracks.

Mel's eyes fought to adjust to the brightness. Then he saw Duke, too.

The boxer had been trotting on the shoulder of the road about two feet away from Jake. The brown grass had hidden his brindle color for a couple of seconds.

The two dogs stood there, sniffing the air as if they were trying to figure out what they should do.

Mary Ellen's voice was startlingly loud in the silence. "It's him, Mel!" she said. "Duke's all right." In a loud voice she called, "Duke! Come, Duke!"

Mel was watching the dogs so intently that he didn't hear her open the door on the other side of the truck. Suddenly she moved to the front of the truck and stopped only when she was standing in the headlights.

"Get back in," Mel commanded in a desperate plea.

"Come, Duke. Good boy," she called.

The distance was still too great for a shot with the tranquilizer gun. Besides, the Doberman seemed much too heavy for the amount the vet had put into the dart. Mel put the gun down gently on the hood of the truck and slowly reached down for the revolver.

"Mary Ellen," he implored her. "Please let me handle this."

She did not respond, but she stood there.

The dogs had not moved. Mel could see that Duke's nubby tail was flicking back and forth, which might be an indication of friendliness. But Mel could not be absolutely sure; the boxer had not responded immediately to his mistress.

"Hey, boy. Come." Mel moved forward into the light

so the dogs could see him. He stooped down, making himself less imposing.

He repeated the soft commands.

There was no response. The only thing Mel knew for sure was that the Doberman's hackles were raised from his pointed nose all the way back to his tail.

"Come, boy," Mel repeated, using the most soothing tone he could muster.

He thought for a moment that the Doberman was beginning to relax slightly, but that was just before Mary Ellen screamed: "No!"

She had seen the gun in Mel's hand.

The dogs lunged forward.

Jake was two leaps on his way before Duke even made a move. But the boxer dug his powerful legs in and reached Jake just as the Doberman was beginning his spring at Mary Ellen.

Duke made a desperate dive at Jake's back at the same instant that Mel managed to get fully erect. The boxer hit the larger dog's hindquarters and spoiled his leap. Both dogs sprawled on the pavement.

"Duke!" Mary Ellen screamed.

Her shrill scream pierced Mel's ears, and he kicked out at the Doberman, who was back on his feet and heading for the girl again. His kick caught Jake in the rump and the dog fell to the side, his snapping jaws missing Mary Ellen by inches. The boxer had also leaped, and when Jake went down Duke's leap carried him right into Mary Ellen's legs. She toppled to the ground.

Mel jumped toward her and kicked at Duke's shoulder.

The boxer spun around and faced Mel.

Mel yelled, hoping to scare the boxer for a second, but instead the dog jumped. Mel twisted his body and Duke flew past, snapping at air where Mel had stood. Mel whirled to see that the Doberman had regained his footing and was going for Mary Ellen, snarling.

Mel brought the revolver up for a shot, but out of the corner of his eye he saw a blur of movement. Duke was heading to intercept the Doberman. But in the flash of the split second, Mel saw he could not reach the girl before the larger dog. Quickly he steadied the gun and squeezed off a shot.

And, in the instant that the gun bucked in his hand, he saw that Duke had been moving faster than he would have believed possible. The bullet caught him in the side and bowled him over.

With the gunshot and Mary Ellen's scream, Jake froze for an instant. He looked back at Mel, and at the twitching body of the boxer. And then he whirled and bolted across the road and into the darkness.

Mel rushed to Mary Ellen and crouched beside her. "Are you okay?" he asked.

She stared at him, and then at the gun in his hand. And then she turned to where Duke was sprawled, still twitching feebly.

"Why?" she said softly. She turned back to Mel, and there was shock and horror on her face. "You promised you'd save him," she cried.

Mel took her by the shoulders and pulled her to her feet. "Stop it," he said in a loud voice. "Just stop."

She shook her head frantically in frustration. "He was only trying to protect me," she cried. "He loved me." And she broke into sobbing.

"Mary Ellen—" he cut himself off. He did not know what to say.

"Leave me alone." She continued to cry. "I want to go home. I want to take Duke home." And she dropped to her knees beside the now still form of the boxer.

Down the road, the headlights were heading their way.

Mel walked out into the field, away from it all.

Instead of saving dogs, he had joined in the killing. And, instead of keeping his promise to Mary Ellen, he had killed her dog—and just when the boxer was trying to help his mistress. It had been an accident, but would the girl ever really believe that, deep down?

Would he?

He stood in the field, listening. Off in the distance he heard a crashing in the undergrowth, and knew it was the Doberman fleeing toward freedom. He smiled bitterly. Probably the only really dangerous dog in the whole pack, and it had gotten away from them all.

He looked back toward the truck, and saw that other cars were pulling up and men were walking around in the headlights, gathering around Mary Ellen and her dead dog. As he watched, one of them helped her up, then picked up the boxer's body and carried it to one of the pickup trucks.

In a moment the truck drove off.

He saw that he had dropped the revolver somewhere in the dark grass. Well, to hell with it. He wouldn't need it again.

Wearily he started back toward the park truck, with the unused tranquilizer gun still lying on its hood.

Epilogue

Mitzie, the little beagle, spent a few days hiding behind a garbage container, then managed to get herself adopted by a couple from McKeesport, Pennsylvania, who were passing through Indiana. She was well loved and she had her puppies, which bore a striking resemblance to smaller versions of Jake. The excess of the litter were dropped off on a quiet road just south of Pittsburgh when Mitzie's new owners could not place them in homes.

Jake? Well, not much was seen of Jake after he escaped the slaughter, but, if you ever see a sturdy-looking Doberman pinscher wandering around your neighborhood, just shout, *"Cell-Bore."* That's it: "Cell-Bore." Those are the words Jake was trained to accept as a command to kill. Guard dogs are not usually trained to the command "Kill," because someone might accidentally trigger a reaction and end up with a sliced throat.

But if you should meet that strange Doberman and give the command, then be ready to defend yourself. Jake has been feral for a long time, but he still remembers some things very well.

Acknowledgments

While this is a work of fiction in the strictest sense, this story could not have been told without the extraordinary assistance of many people who contributed factual information.

The listing of particular individuals or organizations by no means suggests their endorsement or implies their support; rather, it is merely an expression of thanks for the guidance and advice extended.

Initial encouragement and the seed of this story came from W.H.P. Poad, M.R.C.S., L.R.C.P., a physician of unmeasured talents and a friend with undeniable patience.

Veterinarians who saw to our sundry dog problems and kindly submitted to uncountable interrogations are: Gordon Leam, B.V.Sc., M.R.C V.S., Nassau, Bahamas; Richard Thackaberry, D.V.M., Fairfield, Conn.; and Forrest Cole, D.V.M., Frisco, Colorado.

[NB: Any medical/animal errors found herein are to be looked on as errors on the part of the author in translating information into fictional form and not the fault of the original suppliers of that information.]

Organizations that supplied research material or guidance were: Argus Archives, New York City; The

Fund for Animals, New York City; and The Friends of Animals, New York City.

Special, and extensive assistance came from the American Humane Association, Denver, Colorado, and particular encouragement and counsel came from Mr. Don Hegg of the association's Public Relations Office and Roger VanTeyens of its Animal Control Section.

Many sources of academic research and/or scientific papers and published articles supplied an insight into the subject. The major ones were: The Ecology of Feral Dogs in Alabama, by M.D. Scott and Keith Causey; The Life and Times of Shag, A Feral Dog in Baltimore, by Alan M. Beck, ScD; and, Psychology of Pet Ownership (an address delivered before the National Conference on the Ecology of the Surplus Dog and Cat Problem).

Any incidents depicted by the author that may conflict with the findings of these learned men should be considered the by-product of the fiction-writing process.

Additionally, a comment about the locale of this work is in order. Southern Indiana is a beautiful section of that great state, and Brown County State Park is recognized as one of the finest facilities of its kind in the nation. The park is excellently maintained and a wonderful place to enjoy nature at its best. While the physical actions of all incidents in this work are accurately described from authentic cases, no such events have ever happened in Brown County State Park, so far as the author knows. There are no packs of dogs running around in Brown County State Park; but such a thing could happen in almost any community so long as dog owners fail to keep their pets under proper control.

Finally, but certainly not minor in importance, thanks must go to some mutts: Clum, who is no longer with us, as well as Beau and Elmer, who have suffered through the indignities of having their most private behavior patterns scrutinized and their personal experiences depicted in print.

Al Dempsey
Mesa Cortina
Colorado